PROGRESS COACH

New York State Progress Coach,
March-to-March Edition, Mathematics, Grade 4

New York State Progress Coach, March-to-March Edition, Mathematics, Grade 4
248NY ISBN-10: 1-60471-298-8 ISBN-13: 978-1-60471-298-8

Cover Image: The Statue of Liberty. © Jupiter Images

Triumph Learning® 136 Madison Avenue, 7th Floor, New York, NY 10016
Kevin McAliley, President and Chief Executive Officer

© 2009 Triumph Learning, LLC

All rights reserved. No part of this publication may be reproduced in whole or in part, stored in a retrieval system, or transmitted in any form or by any means, electronic, mechanical, photocopying, recording or otherwise, without written permission from the publisher.

Printed in the United States of America.

10 9 8 7 6 5 4 3 2 1

Table of Contents

New York State Indicators

GRADE 3 POST-MARCH LESSONS

Strand 1 — Number Sense and Operations
Lesson 1	Exploring Equivalent Fractions	4	3.N.14
Lesson 2	Comparing and Ordering Unit Fractions on a Number Line	8	3.N.15
Lesson 3	Models for Division	12	3.N.23
Lesson 4	Multiplication and Division Facts	16	3.N.19, 3.N.20, 3.N.22
Lesson 5	Rounding to the Nearest 100	20	3.N.25
Lesson 6	When Is an Estimate Appropriate?	24	3.N.26

Strand 2 — Algebra
Lesson 7	Comparing Fractions Using Symbols	28	3.A.1

Strand 3 — Geometry
Lesson 8	Congruent and Similar Figures	32	3.G.2

Strand 5 — Statistics and Probability
Lesson 9	Collecting and Recording Data in Tables	36	3.S.1, 3.S.2

Progress Check 1 40
Progress Check 2 42

GRADE 4 PRE-MARCH LESSONS

Strand 1 — Number Sense and Operations
Lesson 10	Understanding Place Value	44	4.N.1, 4.N.2, 4.N.4, 4.N.5
Lesson 11	Comparing and Ordering Numbers	48	4.N.3
Lesson 12	The Associative Property of Multiplication	52	4.N.6
Lesson 13	Properties of Multiplying Even and Odd Numbers	56	4.N.13
Lesson 14	Adding and Subtracting Whole Numbers	60	4.N.14
Lesson 15	Models for Multiplication and Division	64	4.N.16
Lesson 16	Relating Multiplication and Division	68	4.N.17
Lesson 17	Multiplying and Dividing by Multiples of 10 and 100	72	4.N.20
Lesson 18	Multiplying Two-Digit Numbers by One-Digit Whole Numbers	76	4.N.18
Lesson 19	Dividing Two-Digit Dividends by One-Digit Divisors	80	4.N.21, 4.N.22
Lesson 20	How to Solve Problems	84	4.N.15
Lesson 21	Rounding to the Nearest Ten and Hundred	88	4.N.26
Lesson 22	Using Estimation to Check Answers	92	4.N.27

Progress Check 1 96
Progress Check 2 98

				New York State Indicators
Strand 2	**Algebra**			
Lesson 23	Using Open Sentences		100	4.A.1
Lesson 24	Comparing Whole Numbers Using Symbols		104	4.A.2, 4.A.3
Lesson 25	Geometric and Number Patterns		108	4.A.4
Lesson 26	Finding the Rule for an Input-Output Table		112	4.A.5
Progress Check 1			116	
Progress Check 2			118	
Strand 3	**Geometry**			
Lesson 27	Two-Dimensional Figures		120	4.G.1, 4.G.2
Lesson 28	Perimeter and Area		124	4.G.3, 4.G.4
Lesson 29	Three-Dimensional Figures		128	4.G.5
Progress Check 1			132	
Progress Check 2			134	
Strand 4	**Measurement**			
Lesson 30	Selecting Tools and Units for Measuring Length		136	4.M.1, 4.M.3
Lesson 31	Measuring Length		140	4.M.2
Lesson 32	Selecting Tools and Units for Measuring Mass		144	4.M.4, 4.M.5
Lesson 33	Selecting Tools and Units for Measuring Capacity		148	4.M.6, 4.M.7
Lesson 34	Making Change		152	4.M.8
Lesson 35	Time		156	4.M.9, 4.M.10
Progress Check 1			160	
Progress Check 2			162	
Strand 5	**Statistics and Probability**			
Lesson 36	Representing Data in Tables, Pictographs, and Bar Graphs		164	4.S.3
Lesson 37	Making Predictions from Data and Graphs		168	4.S.5, 4.S.6
Progress Check 1			172	
Progress Check 2			174	

Progress Test 177
Glossary 197
Punch-Out Tools 201

Lesson 1

Exploring Equivalent Fractions

Math Words to Know

numerator the top number in a fraction, such as the 1 in $\frac{1}{2}$

denominator the bottom number in a fraction, such as the 2 in $\frac{1}{2}$

equivalent fractions different fractions that have equal value, such as $\frac{1}{2}$ and $\frac{2}{4}$

unit fraction a fraction with a numerator of 1, such as $\frac{1}{2}$

Example 1

How much of the square is shaded?

Step 1 Count the number of equal parts.

There are 4 equal parts.

Step 2 Count the number of shaded parts.

There is 1 shaded part.

Step 3 Write $\frac{\text{shaded part}}{\text{total number of parts}}$.

Hint A fraction names part of a whole or a group.

Solution $\frac{1}{4}$ of the square is shaded.

4 Number Sense and Operations

Example 2

The fractions modeled below are equivalent. Write the equivalent fractions for the shaded parts of the two rectangles.

Step 1 Find the number of equal parts in the top rectangle.

There are 6 equal parts.

Step 2 Count the number of shaded parts in the top rectangle.

There are 2 parts shaded.

Step 3 Write $\frac{numerator}{denominator}$.

The rectangle is $\frac{2}{6}$ shaded.

Step 4 Find the fraction of the second rectangle using total parts and shaded parts.

The rectangle is $\frac{1}{3}$ shaded.

Solution $\frac{2}{6}$ and $\frac{1}{3}$ are equivalent fractions.

Multiple-Choice Questions

1. What part of the square is shaded?

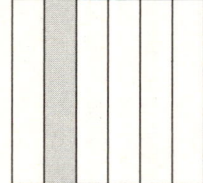

   Test Taking Tip Solve a simpler problem. Count the number of parts.

   A. $\frac{1}{8}$ There are not 8 equal parts. Cross it out.

   B. $\frac{1}{6}$ The square has 6 equal parts. This could be right.

   C. $\frac{1}{2}$ There are not 2 equal parts. Cross it out.

   D. $\frac{5}{6}$ The square has 6 equal parts. This could be right.

2. What part of the circle is shaded?

   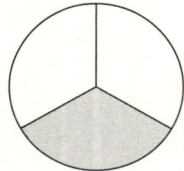

   A. $\frac{1}{3}$

   B. $\frac{1}{2}$

   C. $\frac{2}{3}$

   D. $\frac{3}{1}$

3. What part of the circle is shaded?

   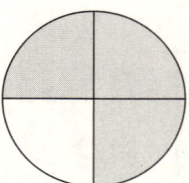

   A. $\frac{1}{4}$

   B. $\frac{2}{3}$

   C. $\frac{3}{4}$

   D. $\frac{3}{1}$

4. Which two fractions name the shaded parts of the rectangles?

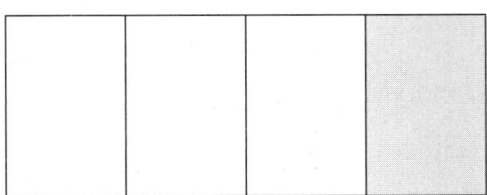

A. $\frac{2}{8}$ and $\frac{1}{4}$ C. $\frac{2}{7}$ and $\frac{1}{4}$

B. $\frac{2}{8}$ and $\frac{1}{8}$ D. $\frac{1}{8}$ and $\frac{1}{4}$

5. Which two fractions name the shaded part of the square?

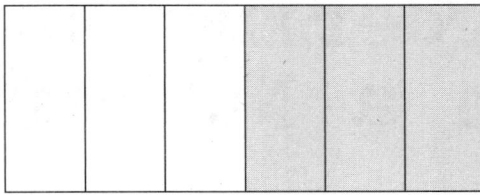

A. $\frac{3}{6}$ and $\frac{2}{5}$

B. $\frac{3}{6}$ and $\frac{3}{4}$

C. $\frac{3}{5}$ and $\frac{1}{2}$

D. $\frac{3}{6}$ and $\frac{1}{2}$

Short-Response Question

6. Ms. Ruiz drew a rectangle on the board.

   Part A   Write a fraction that names the shaded figure.

   Answer _____

   Part B   Write an equivalent fraction that names the shaded figure.

   Answer _____

Lesson 2

Comparing and Ordering Unit Fractions on a Number Line

Math Words to Know

is greater than (>)   an amount that is more than a second amount

is less than (<)   an amount that is less than a second amount

like denominators   fractions with the same denominator, such as $\frac{2}{4}$ and $\frac{3}{4}$

Example 1

Which is greater: $\frac{1}{2}$ or $\frac{1}{3}$?

Step 1   Use a picture to show $\frac{1}{2}$ and $\frac{1}{3}$.

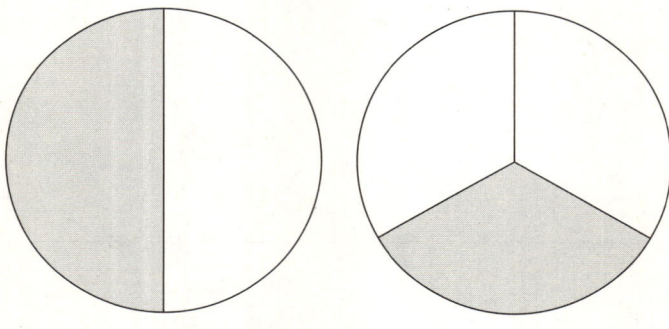

Hint Make your drawings the same size.

Step 2   Compare the sections that are shaded.

The circle with $\frac{1}{2}$ has the larger section shaded.

Solution $\frac{1}{2}$ is greater than $\frac{1}{3}$.

8   Number Sense and Operations

Example 2

Which is greater: $\frac{1}{4}$ or $\frac{3}{4}$?

Step 1 Make a number line from 0 to 1. Divide it into fourths.

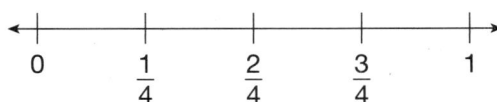

Step 2 Put a mark at $\frac{1}{4}$.

Put a mark at $\frac{3}{4}$.

Find which fraction is farther to the right.

Hint If the denominators are the same, the fraction with the greater numerator is greater.

Solution $\frac{3}{4}$ is greater than $\frac{1}{4}$.

Example 3

Write the fractions $\frac{1}{2}$, $\frac{1}{4}$, and $\frac{1}{3}$ from least to greatest.

Step 1 Make a number line for each unit fraction.

Step 2 Put marks at $\frac{1}{2}$, $\frac{1}{4}$, and $\frac{1}{3}$ on the number lines.

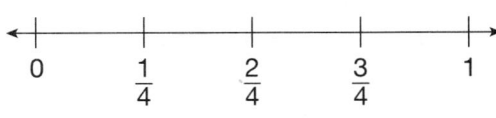

Step 3 Compare the marks.

Hint The fraction farthest from 0 is the greatest fraction.

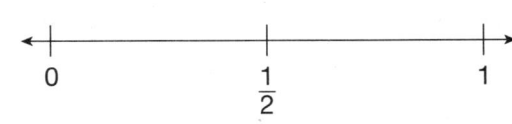

$\frac{1}{4}$ is the least fraction.

$\frac{1}{3}$ is the next greatest fraction.

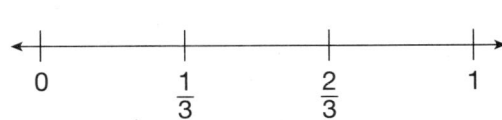

$\frac{1}{2}$ is the greatest fraction.

Solution $\frac{1}{4}, \frac{1}{3}, \frac{1}{2}$

Multiple-Choice Questions

1. Which sentence is true?

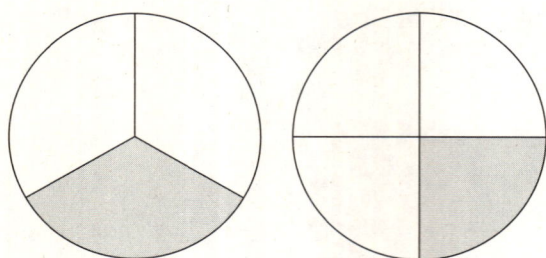

A. $\frac{1}{4}$ is greater than $\frac{1}{3}$.

B. $\frac{1}{3}$ is equal to $\frac{1}{4}$.

C. $\frac{1}{3}$ is less than $\frac{1}{4}$.

D. $\frac{1}{4}$ is less than $\frac{1}{3}$.

Test Taking Tip Use the drawing to help you compare.

Is the shaded section of the second circle bigger than the shaded section of the first circle?

Are the shaded sections equal in size?

Is the shaded section of the first circle smaller than the shaded section of the second circle?

Is the shaded section of the second circle smaller than the shaded section of the first circle?

2. Which sentence is true?

A. $\frac{1}{3}$ is greater than $\frac{1}{2}$.

B. $\frac{1}{2}$ is greater than $\frac{1}{3}$.

C. $\frac{1}{2}$ is less than $\frac{1}{3}$.

D. $\frac{1}{2}$ is equal to $\frac{1}{3}$.

3. Which sentence is true?

A. $\frac{1}{4}$ is greater than $\frac{1}{2}$.

B. $\frac{1}{2}$ is less than $\frac{1}{4}$.

C. $\frac{1}{2}$ is greater than $\frac{1}{4}$.

D. $\frac{1}{2}$ is equal to $\frac{1}{4}$.

4. Which sentence is true?

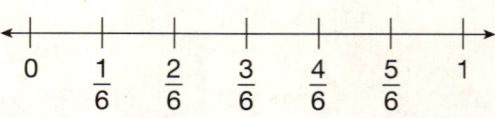

A. $\frac{1}{6}$ is greater than $\frac{3}{6}$.

B. $\frac{4}{6}$ is greater than $\frac{2}{6}$.

C. $\frac{5}{6}$ is less than $\frac{3}{6}$.

D. $\frac{4}{6}$ is less than $\frac{1}{6}$.

5. Which sentence is true?

   A. $\frac{5}{8}$ is greater than $\frac{4}{8}$.

   B. $\frac{3}{8}$ is greater than $\frac{7}{8}$.

   C. $\frac{6}{8}$ is less than $\frac{5}{8}$.

   D. $\frac{2}{8}$ is less than $\frac{1}{8}$.

6. Which sentence is true?

   A. $\frac{3}{10}$ is less than $\frac{1}{10}$.

   B. $\frac{9}{10}$ is greater than $\frac{5}{10}$.

   C. $\frac{3}{10}$ is less than $\frac{1}{10}$.

   D. $\frac{6}{10}$ is greater than $\frac{9}{10}$.

7. Which shows the fractions in order from greatest to least?

   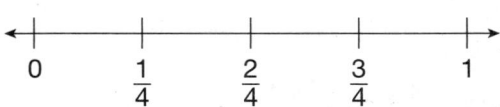

   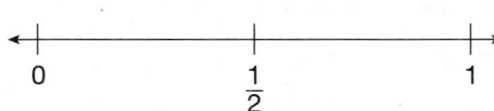

   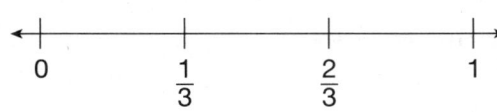

   A. $\frac{1}{2}, \frac{1}{3}, \frac{1}{4}$

   B. $\frac{1}{3}, \frac{1}{2}, \frac{1}{4}$

   C. $\frac{1}{4}, \frac{1}{3}, \frac{1}{2}$

   D. $\frac{1}{2}, \frac{1}{4}, \frac{1}{3}$

Short-Response Question

8. Ray ate $\frac{1}{3}$ of a sandwich.

   Billy ate $\frac{1}{2}$ of a sandwich of the same size.

   Allie ate $\frac{1}{5}$ of a sandwich of the same size.

   Part A Order the fractions from least to greatest.

   Answer _____

   Part B Who ate the greatest part of a sandwich?

   Answer _____

Lesson 3
Models for Division

Math Words to Know

array pictures, numbers, or objects arranged in rows and columns

division separation of numbers into smaller, equal groups

dividend in division, the number that is divided

divisor in division, the number that divides the dividend

quotient in division, the answer to the problem

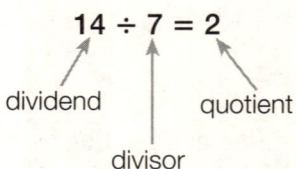

Example 1

Write a division fact for the array below.

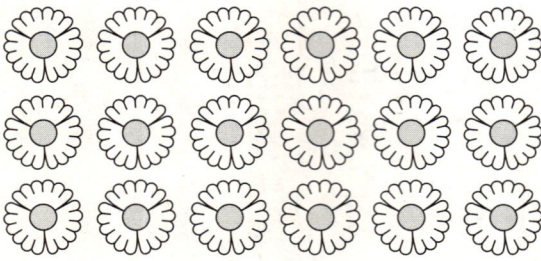

Step 1 Count the number of flowers. This is the dividend.

There are 18 flowers in all. 18 is the dividend.

Step 2 Count the number of columns. This is the divisor.

There are 6 columns. 6 is the divisor.

Step 3 Count the number of rows. This is the quotient.

There are 3 rows. 3 is the quotient.

Step 4 Use the numbers to write a division sentence.

Hint Use: dividend ÷ divisor = quotient

Solution 18 ÷ 6 = 3

Example 2

Use repeated subtraction to show that 15 ÷ 3 = 5.

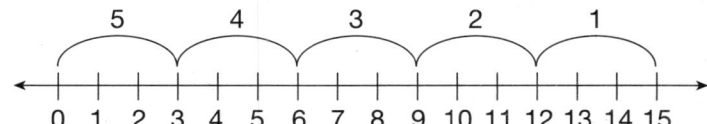

Step 1 Start at 15 and subtract groups of 3s until you reach 0. Use your finger to trace the groups.

You subtracted 3 five times.

Step 2 Write the subtraction sentence five times.

Start with 15 − 3 = 12.

Solution 15 − 3 = 12

12 − 3 = 9

9 − 3 = 6

6 − 3 = 3

3 − 3 = 0

Hint Division makes equal groups. Repeated subtraction also makes equal groups.

Multiple-Choice Questions

1. Which division fact is shown by the array?

		Test Taking Tip Count the number of books in the array.
A.	20 ÷ 5 = 4	There are 20 books. This could be the answer.
B.	20 ÷ 4 = 4	There are 20 books. This could be the answer.
C.	10 ÷ 5 = 2	There are 10 books. This cannot be the answer.
D.	5 ÷ 4 = 20	There are 5 books. This cannot be the answer.

2. Which array shows 8 ÷ 4 = 2?

   A. ☆☆☆☆☆☆☆☆

   C.

   B.

   D. ☆☆
      ☆☆
      ☆☆
      ☆☆
      ☆☆

14   Number Sense and Operations

3. Which number sentence is shown by the array?

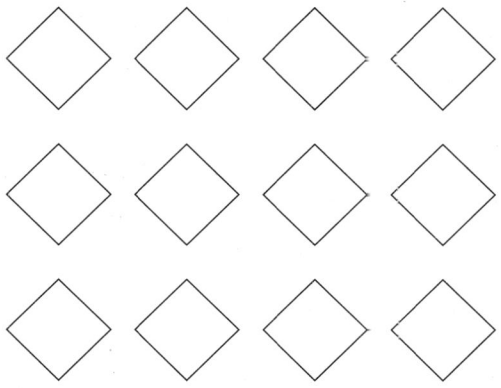

A. $4 \div 3 = 12$

B. $12 \div 4 = 3$

C. $14 \div 2 = 2$

D. $12 \div 1 = 12$

4. The number line shows repeated subtraction. Which division sentence is shown?

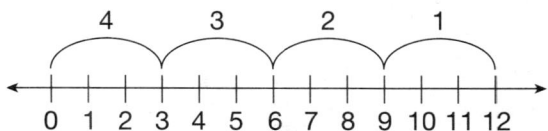

A. $12 \div 2 = 6$

B. $12 \div 3 = 4$

C. $4 \div 1 = 3$

D. $4 \div 3 = 1$

Short-Response Question

5. Andre made an array of stars. It shows this division sentence:

   $15 \div 5 = 3$

   Part A   Draw the array.

   Part B   Label the dividend, divisor, and quotient in your drawing.

Lesson 3: Models for Division   15

Lesson 4

Multiplication and Division Facts

Math Words to Know

multiplication table a table of multiplication facts

row a line in a table or an array that goes across

column a line in a table or an array that goes up and down

product the answer to a multiplication problem

Example 1

Fill in row 3 on the multiplication table.

×	0	1	2	3	4	5	6	7	8	9	10
0	0	0	0	0	0	0	0	0	0	0	0
1	0	1	2	3	4	5	6	7	8	9	10
2	0	2	4	6	8	10	12	14	16	18	20
3											
4	0	4	8	12	16	20	24	28	32	36	40
5	0	5	10	15	20	25	30	35	40	45	50
6											
7											
8	0	8	16	24	32	40	48	56	64	72	80
9											
10	0	10	20	30	40	50	60	70	80	90	100

Step 1 Find row 3.

Start with 0. Write **0**.

Add 3 and fill in the next box, **3**.

Step 2 Continue adding 3 until you have filled all the squares in row 3.

Hint The 3 row and the 3 column are the same. Use the 3 column to check your answers.

Solution Row 3 numbers are 0, 3, 6, 9, 12, 15, 18, 21, 24, 27, and 30.

16   Number Sense and Operations

Example 2

7 × 5 = _____

    Step 1 Find row 7.

               Put a finger on the 7.

               Move your finger across row 7 until it is on the 5 column.

    Step 2 Look at the number.

               The number is 35.

Solution 7 × 5 = 35.

Example 3

A package has 4 notebooks. Claudia bought 3 packages. How many notebooks did she buy?

3 × 4 = _____

    Step 1 Use the multiplication table.

               Find row 4.

               Put your ringer on the 4.

               Move your finger across row 4 until it is on the 3 column.

    Step 2 Look at the number.

               The number is 12.

               3 × 4 = 12

Solution Claudia bought 12 notebooks.

Multiple-Choice Questions

1. 8 × 4 = _____

   Test Taking Tip Use the multiplication table. Find row 8 and column 4. Where do they meet?

   A. 10   This number is not enough. Cross it out.

   B. 12   This number is in the 4 column. This could be the answer.

   C. 17   This number is not enough. Cross it out.

   D. 32   This number is in row 8 and the 4 column. This could be the answer.

2. 9 × 7 = _____

   A. 7
   B. 16
   C. 63
   D. 81

3. Justin walked 4 miles each day for 5 days. How many miles did he walk?

   4 × 5 = _____

   A. 11
   B. 20
   C. 22
   D. 31

4. 4 × 4 = _____

   A. 8
   B. 16
   C. 28
   D. 54

5. Emiko works 5 hours a day. She works 5 days. How many hours does she work in all?

   5 × 5 = _____

   A. 25
   B. 29
   C. 30
   D. 32

6. In the multiplication table, what is the pattern in row 6?

   A. multiply 6

   B. subtract 6

   C. divide 6

   D. add 6

7. 6 × 6 = _____

   A. 12

   B. 18

   C. 36

   D. 38

8. What is the pattern in row 3 of the multiplication table?

   A. add 1

   B. add 3

   C. multiply by 3

   D. multiply by 6

9. 10 × 8 = _____

   A. 80

   B. 82

   C. 86

   D. 90

Short-Response Question

10. Use the multiplication table.

    Part A Find the answer to the multiplication problem 8 × 6.

    Write how you found the product.

    _____

    _____

    _____

    Part B The product of 5 × 4 can be found in two places on the table.

    Write where you found the product.

    _____

    _____

    _____

Lesson 5

Rounding to the Nearest 100

Math Words to Know

rounding to estimate the value of a number based on a given place value

Rounding to the nearest ten, 2<u>4</u>8 rounds up to 250.

Rounding to the nearest hundred, <u>2</u>48 rounds down to 200.

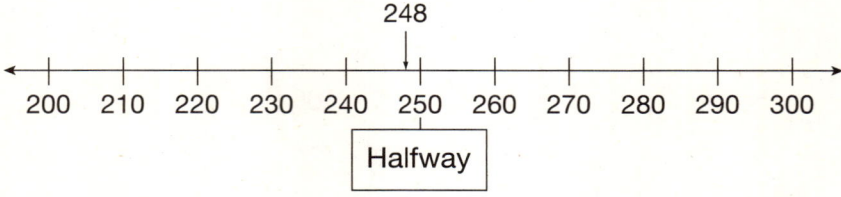

Example 1

Round 237 to the nearest 10.

Step 1 Find the tens place. Underline it.

2<u>3</u>7

Step 2 Circle the digit to the right of the tens place.

2<u>3</u>⑦

Step 3 The ones digit is 7.

The tens digit rounds up.

The new tens digit is 4.

Hint If the ones digit is less than 5, round down. If the ones digit is 5 or greater, round up.

Step 4 Change the ones digit to 0.

Solution 237 rounded to the nearest 10 is 240.

Example 2

Round 237 to the nearest 100.

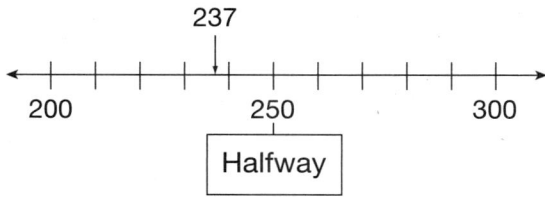

Step 1 Underline the hundreds place.

2̲37

Step 2 Circle the digit to the right of the hundreds place.

2̲③7

Step 3 The tens digit is less than 5.

Round down.

The underlined digit stays the same.

Step 4 Change each digit to the right of the hundreds place to 0.

Solution 237 rounded to the nearest 100 is 200.

Multiple-Choice Questions

1. Round 42 to the nearest 10.

   Test Taking Tip Look to the right of the tens digit. Then round up or round down.

   A. 30   The tens place is rounded down too much. Cross it out.

   B. 40   The ones place is rounded down. This could be the answer.

   C. 45   45 is not the nearest 10. Cross it out.

   D. 50   The tens place is rounded up. Cross it out.

2. Round 364 to the nearest 100.

   A. 300
   B. 360
   C. 400
   D. 470

3. To the nearest 10, which number does **not** round to 70?

   A. 64
   B. 65
   C. 68
   D. 72

4. To the nearest 100, which number does **not** round to 100?

   A. 120
   B. 130
   C. 140
   D. 160

5. 250 is halfway between which numbers?

   A. 100 and 200
   B. 200 and 300
   C. 300 and 400
   D. 400 and 500

6. Round 46 to the nearest 10.
   A. 30
   B. 40
   C. 50
   D. 70

7. Round 324 to the nearest 100.
   A. 100
   B. 200
   C. 300
   D. 400

8. Round 850 to the nearest 100.
   A. 700
   B. 800
   C. 850
   D. 900

9. Round 473 to the nearest 100.
   A. 400
   B. 470
   C. 480
   D. 500

Short-Response Question

10. Jordan has 168 baseball cards.

    Part A Round 168 to the nearest 10.

    Answer _____

    Part B Round 168 to the nearest 100.

    Answer _____

Lesson 6

When Is an Estimate Appropriate?

Math Words to Know

estimate   a number that is close to the exact amount

Example 1

Rusty's grandparents live 198 miles away from his home. His cousin lives 125 miles away from his home. About how much farther does Rusty live from his grandparents than from his cousin?

Step 1   Round 198 and 125 to the nearest hundred using number lines.

Hint The word "about" in a question is a clue word that you can estimate.

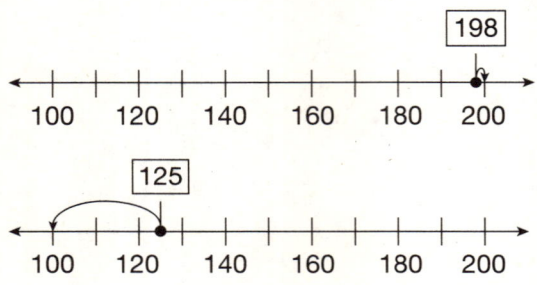

Step 2   Subtract using the rounded numbers.

```
 198  ⟶  rounds up to    ⟶    200
- 125  ⟶  rounds down to  ⟶   - 100
                                ___
                                100
```

Solution   Rusty lives about 100 miles farther from his grandparents than from his cousin.

24   Number Sense and Operations

Example 2

A female black bear weighs 337 pounds. A male black bear weighs 156 pounds more than the female bear. How much does the male black bear weigh?

Write the exact number and an estimate.

Step 1 Decide which operation you will use.

   This problem requires addition.

Step 2 Find the exact answer. How much does the male bear weigh?

```
   337
 + 156
 -----
   493
```

The male bear weighs 493 pounds.

Step 3 Find the estimate. Round the numbers to nearest hundred. Then add.

Hint Decide what place you should round each number to before you begin working.

```
   337   ——▶  rounds down to  ——▶     300
 + 156   ——▶  rounds up to    ——▶   + 200
 -----                               -----
   493                                 500
```

The male bear weighs about 500 pounds.

Solution 493 pounds is the exact answer. 500 pounds is the estimate.

Multiple-Choice Questions

1. Ken has 78 toy cars. He gives Abe 24 of them.

   About how many toy cars does Ken have left?

   Test Taking Tip Round both numbers to the nearest ten. Then subtract.

   A. 20   This answer is too low. Cross it out.

   B. 30   This answer is too low. Cross it out.

   C. 50   This could be the answer.

   D. 60   This could be the answer.

Use the table below for questions 2 and 3.

Garden Sale

Item	Number Sold
Plants	223
Trees	335
Flowers	107

2. About how many more trees than plants were sold?

   A. 100   C. 200

   B. 150   D. 558

3. Exactly how many trees and flowers were sold?

   A. 228

   B. 400

   C. 442

   D. 500

4. The kite shop has 74 kites. It sells 39 kites. Exactly how many kites are left?

   A. 30

   B. 35

   C. 70

   D. 110

5. The Dover family drove 465 miles in three days. Then they drove 105 miles the next day. About how many miles did the family drive in all?

   A. 300

   B. 400

   C. 500

   D. 600

6. Amy picked 24 red apples and 35 green apples. About how many apples did she pick in all?

   A. 20 C. 60
   B. 50 D. 70

7. Christine has $20. She spent $8 on dinner. Exactly how much money does she have left?

   A. $10 C. $13
   B. $12 D. $14

8. Randy bowled a game and scored 138 points. Rochelle bowled a game and scored 150 points. Exactly how many points did they score in all?

   A. 12 C. 288
   B. 200 D. 300

9. A quart of milk is 32 ounces. About how many ounces in 2 quarts?

   A. 30 C. 50
   B. 40 D. 60

Short-Response Question

10. Josh read 29 books and Noah read 22 books.

    Part A About how many total books did they read?

    Show your work.

    Answer _____

    Part B Exactly how many total books did they read?

    Answer _____

Lesson 6: When Is an Estimate Appropriate?   27

Lesson 7

Comparing Fractions Using Symbols

Math Words to Know

is greater than (>) an amount that is more than a second amount

is less than (<) an amount that is less than a second amount

unit fraction a fraction with a numerator of 1, such as $\frac{1}{2}$

Example 1

What symbol will make this sentence true?

$$\frac{1}{6} \bigcirc \frac{1}{2}$$

Step 1 Show $\frac{1}{6}$ and $\frac{1}{2}$ using fraction strips.

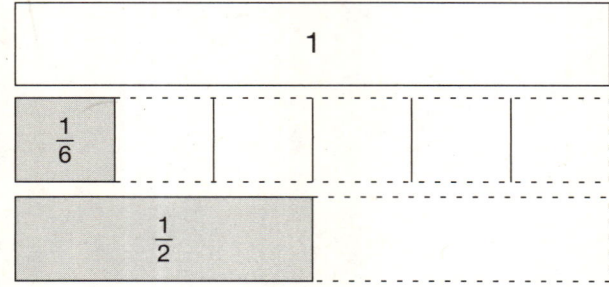

Hint With unit fractions, the fraction with the least digit for the denominator is the greater fraction.

Step 2 Compare the fraction strips.

The shaded section for $\frac{1}{6}$ is smaller than the shaded section for $\frac{1}{2}$.

Solution $\frac{1}{6} < \frac{1}{2}$

Example 2

What symbol will make this sentence true?

$$\frac{1}{4} \bigcirc \frac{1}{2}$$

Step 1 Use the number lines.

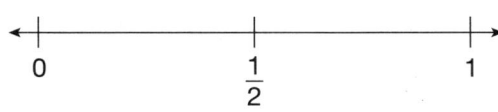

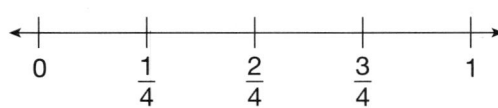

Step 2 Find $\frac{1}{2}$ on the top number line. Find $\frac{1}{4}$ on the bottom number line.

Step 3 Find the fraction that is farther to the right.

$\frac{1}{2}$ is farther to the right than $\frac{1}{4}$. It is the greater fraction.

Solution $\frac{1}{4} < \frac{1}{2}$

Multiple-Choice Questions

For questions 1–3, use the number lines.

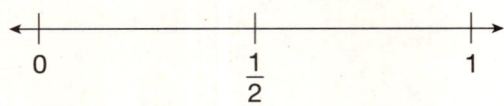

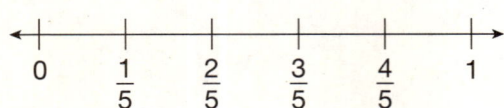

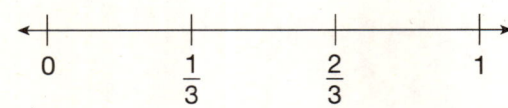

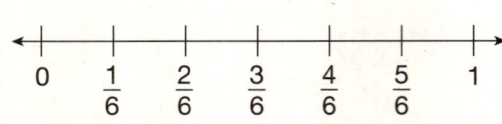

1. Which symbol will make this sentence true?

    $\frac{1}{2} \bigcirc \frac{1}{6}$

    Test Taking Tip Cross out the answer choices that do not make sense.

    A. < This is a comparing symbol. This could be right.

    B. − This is not a comparing symbol. Cross it out.

    C. > This is a comparing symbol. This could be right.

    D. + This is not a comparing symbol. Cross it out.

2. Which symbol will make this sentence true?

    $\frac{1}{3} \bigcirc \frac{1}{5}$

    A. =   C. <
    B. +   D. >

3. Which symbol will make this sentence true?

    $\frac{1}{5} \bigcirc \frac{1}{2}$

    A. <   C. =
    B. >   D. +

30   Algebra

4. Which symbol will make this sentence true?

$\frac{1}{4} \bigcirc \frac{1}{3}$

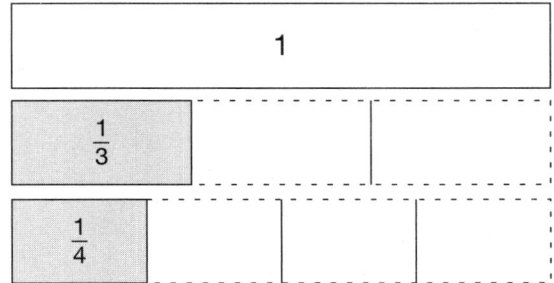

A. =
B. <
C. >
D. +

5. Which symbol will make this sentence true?

$\frac{1}{4} \bigcirc \frac{1}{10}$

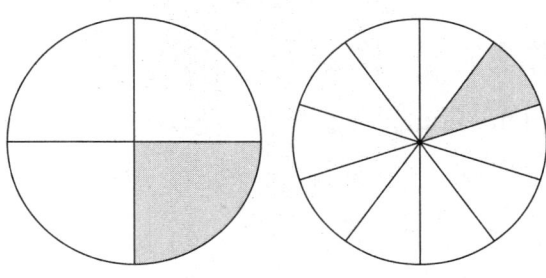

A. =
B. <
C. >
D. +

Short-Response Question

6. Ms. Simmons asked her class to draw fraction strips for $\frac{1}{2}$ and $\frac{1}{4}$.

Part A   Draw the fraction strips.

Part B   Which symbol will make this sentence true?

$\frac{1}{2} \bigcirc \frac{1}{4}$

Answer _____

Lesson 7: Comparing Fractions Using Symbols

Lesson 8

Congruent and Similar Figures

Math Words to Know

plane figure   a flat, two-dimensional figure like a square, circle, or rectangle

congruent figures   plane figures that have the same size and the same shape

similar figures   plane figures that have the same shape, but not necessarily the same size

Example 1

Which figure is congruent to A?

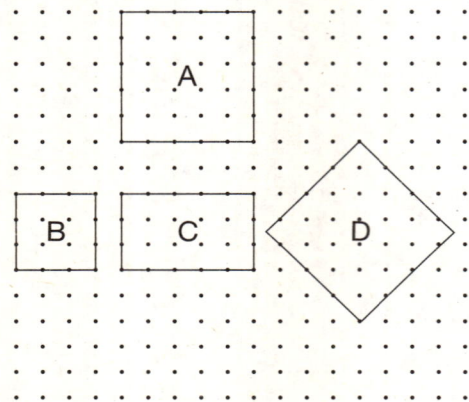

Step 1   Look at the figures. Which figures have the same shape as A?

B and D have the same shape as figure A.

Step 2   Which figures are the same size as A?

D is the same size as A.

Hint   The position of the figure doesn't matter.

Solution   D is the same shape and size as A. D is congruent to A.

32   Geometry

Example 2

Which figure is not similar to A?

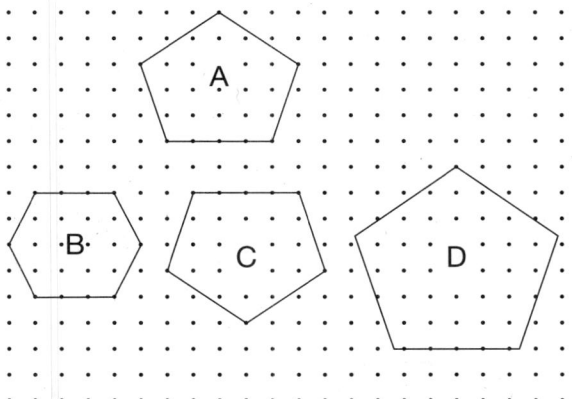

Step 1 Look at the figures. Which figures have the same shape as A?

C and D are the same shape as A.

Hint Similar figures can also be the same size.

Step 2 Which figure is not the same shape as A?

B is not the same shape as A. It has 6 sides.

Solution B is not similar to A.

Multiple-Choice Questions

1. Which figure is congruent to the triangle?

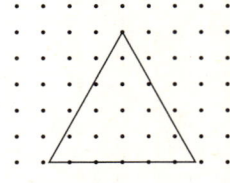

Test Taking Tip Look at all the answer choices. Cross out the answers that are not the same shape and size as the triangle.

A.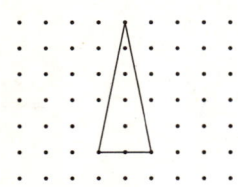

This is not the same shape. Cross it out.

B.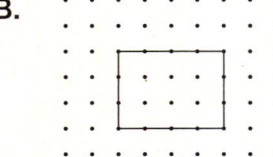

This is not the same shape. Cross it out.

C.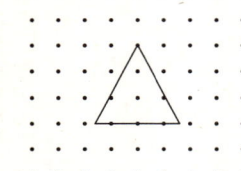

This is the same shape. This could be right.

D.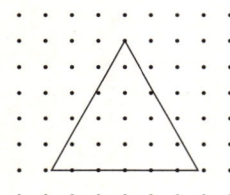

This is the same shape. This could be right.

2. Which pair are congruent shapes?

A. 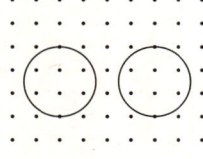   B. 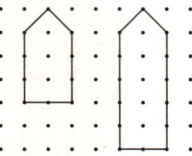   C. 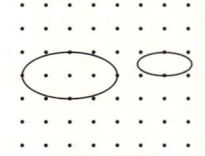   D.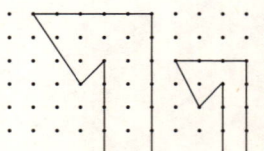

3. Which figure is similar to the rectangle?

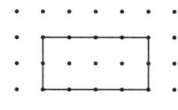

A.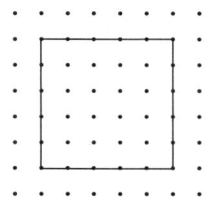

C.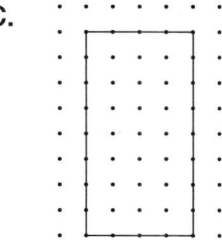

B.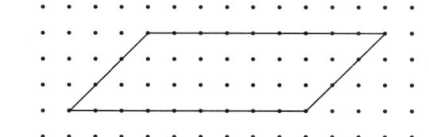

D.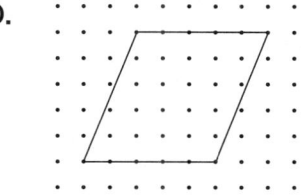

Short-Response Question

4. Look at the rhombus.

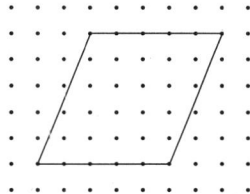

Part A Draw a rhombus that is congruent to the figure above.

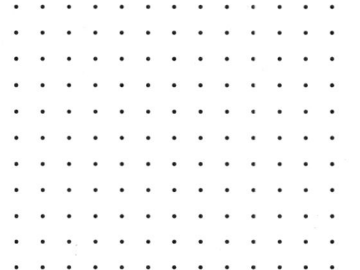

Part B Draw a rhombus that is similar but **not** congruent to the figure above.

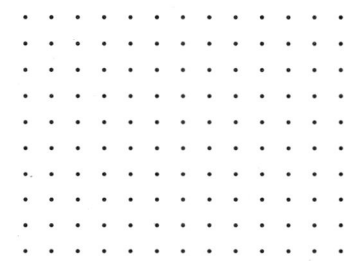

Lesson 8: Congruent and Similar Figures   35

Lesson 9

Collecting and Recording Data in Tables

Math Words to Know

data   a set of collected information

survey   information about a group that is found by questioning or observing

tally table   a table that shows data by using a tally mark for each item counted

Example 1

Tomas asked classmates how they get to school. His results are shown in the table.

How Do You Get to School?

Method	Number of Students
Bus	48
Car	16
Walking	7
Bicycle	3

How many more students ride the bus than walk to school?

Step 1   Find how many students ride the bus to school.

   48 students ride the bus.

Step 2   Find how many students walk to school.

   7 students walk.

Step 3   Find how many more students ride the bus than walk to school.

   Subtract 48 − 7.

Solution   41 more students ride the bus than walk to school.

36   Statistics and Probability

Example 2

Stephanie asked people to name their favorite type of movie. She made the tally table below.

Favorite Type of Movie

Type of Movie	Tally
Action/Adventure	|||| ||
Romance	|||| ||||	
Comedy	||||	
Sports		

How many people chose comedy?

Step 1 Add a third column to the table.

Count the tally marks for each movie type.

Hint Each | counts as 1.
Each |||| counts as 5.

Record the number in the new column.

Favorite Type of Movie

Type of Movie	Tally	Number
Action/Adventure	|||| ||	7
Romance	|||| ||||		11
Comedy	||||		6
Sports			1

Step 2 There are 6 tally marks for comedy.

Solution 6 people chose comedies as their favorite type of movie.

Lesson 9: Collecting and Recording Data in Tables

Multiple-Choice Questions

1. The table shows results from a survey about people's favorite video game.

    Favorite Video Game

Video Game	Number
Space Attack	36
Baseball	13
Quest	24
Card Games	9

    How many more people like Quest than Baseball?

    Test Taking Tip Estimate to see if your answer makes sense.

A.	9 people	If I round, I get 20 − 10 = 10. This could be right.
B.	11 people	If I round, I get 20 − 10 = 10. This could be right.
C.	19 people	If I round, I get 20 − 10 = 10. This number is too great. Cross it out.
D.	22 people	If I round, I get 20 − 10 = 10. This number is too great. Cross it out.

    ---

2. Brian made a table of students' hair color.

    Students' Hair Color

Hair Color	Number
Brown	18
Black	19
Red	4
Blond	11

    How many more students have black hair than blond hair?

    A. 1 students C. 8 students

    B. 7 students D. 15 students

3. Danielle asked people what month they were born. She made the tally table below.

    Birth Month

Month	Tally
January	|||| |||
March	||||
July	|||| ||||
September	|||| |||
December	|||| ||

    How many people were born in September?

    A. 4 people C. 8 people

    B. 7 people D. 10 people

38 Statistics and Probability

4. Marcus asked his friends how many books they read over the summer. He made the tally table below.

Number of Books Read over the Summer

Number of Books	Tally
3	⋕ I
4	IIII
5	III
6	II

How many of Marcus's friends read more than 4 books?

A. 3 friends

B. 4 friends

C. 5 friends

D. 9 friends

5. Brenda asked students what their favorite vegetable is. Her results are shown below.

What is your favorite vegetable?
broccoli	corn	carrots
corn	carrots	broccoli
corn	corn	carrots
corn	carrots	corn
corn		broccoli

Which tally shows the number of students who like corn the most?

A. IIII

B. ⋕ I

C. ⋕ II

D. ⋕ IIII

Short-Response Question

6. Anthony surveyed people on their favorite type of cookie. His results are shown below.

What is your favorite type of cookie?
chocolate chip	oatmeal
peanut butter	chocolate chip
chocolate chip	oatmeal
chocolate chip	chocolate chip
oatmeal	oatmeal
chocolate chip	chocolate chip
peanut butter	oatmeal
chocolate chip	peanut butter

Make a tally table to show the data.

Lesson 9: Collecting and Recording Data in Tables 39

PROGRESS CHECK 1

1 Which number sentence is shown by the array?

A $20 \div 4 = 5$ C $20 \div 5 = 5$
B $5 \div 4 = 20$ D $20 \div 4 = 20$

2 Which figure is congruent to the parallelogram?

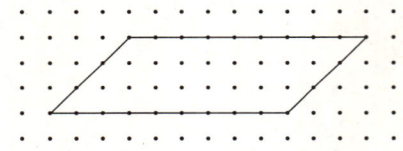

F

G

H

J

3 Round 539 to the nearest 100.

A 400   C 540
B 500   D 600

4 Which symbol will make this sentence true?

$$\frac{1}{4} \underline{} \frac{1}{10}$$

F >
G =
H <
J +

5 $8 \times 4 =$ _____

A 12
B 28
C 32
D 36

6 Tamara surveyed her friends on their favorite type of music. Her results are shown in the tally table.

Favorite Type of Music

Type of Music	Tally
Dance	ЖЖ I
Rock/Pop	IIII
Hip-Hop	ЖЖ III
Country	II

How many more of Tamara's friends like hip-hop than country?

F 3   H 5
G 4   J 6

7 What fraction of the rectangle is shaded?

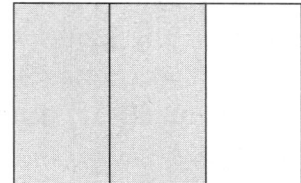

A $\frac{1}{3}$ C $\frac{2}{3}$

B $\frac{1}{2}$ D $\frac{3}{2}$

Short-Response Question

8 Look at the shapes below.

Part A

Write a division sentence that is shown by the array.

Answer _____

Part B

Explain how you found your answer for Part A.

Progress Check 1   41

PROGRESS CHECK 2

1 Jasmine bought 3 packs of gum. Each pack of gum has 9 pieces. How many pieces of gum did Jasmine buy?

A 12 pieces
B 18 pieces
C 21 pieces
D 27 pieces

2 Which symbol will make this sentence true?

$$\frac{1}{6} \underline{\quad} \frac{1}{5}$$

F >
G <
H +
J =

3 Derrick has 64 seashells. He found 18 more. About how many seashells does Derrick have now?

A 40
B 70
C 80
D 90

4 Which shows the fractions in order from least to greatest?

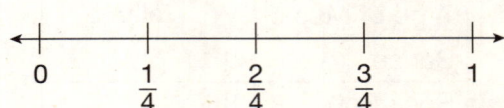

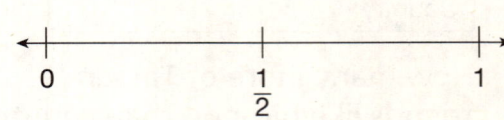

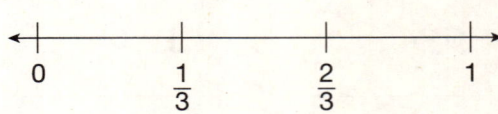

F $\frac{1}{2}, \frac{1}{3}, \frac{1}{4}$ H $\frac{1}{4}, \frac{1}{3}, \frac{1}{2}$

G $\frac{1}{3}, \frac{1}{2}, \frac{1}{4}$ J $\frac{1}{2}, \frac{1}{4}, \frac{1}{3}$

5 What are two ways to name the shaded part of the square?

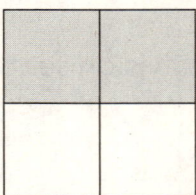

A $\frac{1}{3}$ and $\frac{2}{6}$ C $\frac{3}{6}$ and $\frac{3}{4}$

B $\frac{1}{4}$ and $\frac{2}{6}$ D $\frac{1}{2}$ and $\frac{2}{4}$

6 In which should you use an exact amount?

F finding how much change you should get back after buying an item

G measuring the amount of food you feed your dog each week

H finding the number of miles between two distant cities

J figuring out how long you slept last night

7 About **800** people were at a dog show. Which could **not** be the exact number of people at the show?

A 754 C 829
B 804 D 860

Short-Response Question

8 Natalie surveyed people on their favorite type of juice. Her results are shown below.

What is your favorite type of juice?	
orange	cranberry
orange	apple
cranberry	cranberry
grape	apple
apple	grape
cranberry	cranberry
cranberry	apple
apple	orange

Make a tally table to show the data.

Favorite Type of Juice

Type of Juice	Tally

Lesson 10

Understanding Place Value

Math Words to Know

place value   the value of a digit in a number

standard form   a number written with only numerals, such as 3,762

expanded form   a way of writing a number as the sum of the values of its digits; 3,762 in expanded form is 3,000 + 700 + 60 + 2

Example 1

What is the value of 6 in 6,091?

Step 1   Make a place-value chart to show the value of each digit.

Thousands	Hundreds	Tens	Ones
6 ,	0	9	1

Step 2   Find the 6 in the chart.

6 is in the thousands place.

Step 3   Multiply to find the value.

6 × 1,000 = 6,000

Solution   The value of 6 in 6,091 is 6,000.

Example 2

Write 8,237 in expanded form.

Step 1 Make a place-value chart to show the value of each digit.

Thousands	Hundreds	Tens	Ones
8 ,	2	3	7

Step 2 Use the chart to find the value of each digit.

There are 8 thousands, or 8,000.

There are 2 hundreds, or 200.

There are 3 tens, or 30.

There are 7 ones, or 7.

Step 3 Write the values as a sum.

Solution 8,237 in expanded form is 8,000 + 200 + 30 + 7.

Example 3

Todd's favorite baseball player had 2,042 career hits. Write the number using words.

Step 1 Find the value of each digit.

There are 2 thousands.

There are 0 hundreds.

There are 4 tens.

There are 2 ones.

Hint When a place value is zero, do not write the value with words.

Step 2 Write each digit with words.

2 thousands becomes "two thousand."

4 tens becomes "forty."

2 ones becomes "two."

Solution The number 2,042 written in words is two thousand, forty-two.

Multiple-Choice Questions

1. Which number is shown in the place-value chart?

Thousands	Hundreds	Tens	Ones
7 ,	6	9	0

Test Taking Tip Count the number of digits in the place value chart.

- A. 769 — This number has 3 digits. Cross it out.
- B. 7,690 — This number has 4 digits. This could be right.
- C. 7,609 — This number has 4 digits. This could be right.
- D. 70,690 — This number has 5 digits. Cross it out.

2. What is the value of 7 in the number 1,784?

   A. 7 ones
   B. 7 tens
   C. 7 hundreds
   D. 7 thousands

3. What is the value of 4 in the number 4,039?

   A. 4
   B. 40
   C. 400
   D. 4,000

4. Which of the following is the expanded form of 9,125?

   A. 9,000 + 120 + 5
   B. 9,000 + 100 + 25
   C. 9,000 + 100 + 20 + 5
   D. 90,000 + 100 + 20 + 5

5. Which of the following is the expanded form of 3,804?

   A. 3,000 + 84
   B. 3,000 + 800 + 4
   C. 3,000 + 80 + 4
   D. 3,000 + 800 + 40

6. Which number has this expanded form?

    6,000 + 50 + 2

    A. 652
    B. 6,052
    C. 6,502
    D. 60,052

7. There were 3,467 people at a concert. What is this number in word form?

    A. three hundred, four hundred sixty-seven
    B. three, four hundred sixty-seven
    C. three thousand, four sixty-seven
    D. three thousand, four hundred sixty-seven

8. Which is nine thousand, seven hundred two?

    A. 972
    B. 9,072
    C. 9,702
    D. 9,720

9. Sal's Pizza Place served 1,083 slices of pizza last weekend. What is this number in word form?

    A. one thousand, eighty-three
    B. one thousand, eight hundred three
    C. one thousand, eight hundred thirty
    D. one hundred eighty-three

Short-Response Questions

10. Write 3,591 in expanded form.

    _____

11. The Key Club raised $2,605 for charity.

    Part A What is the value of the 6 in the number?

    _____

    Part B Write the number in word form.

    _____ dollars

Lesson 10: Understanding Place Value   47

Lesson 11

Comparing and Ordering Numbers

Math Words to Know

is greater than (>)   an amount that is more than a second amount

is less than (<)   an amount that is less than a second amount

Example 1

Which symbol makes this number sentence true?

2,561 _____ 2,651

Step 1   Make a place-value chart to show the value of the digits in each number.

Thousands		Hundreds	Tens	Ones
2	,	5	6	1
2	,	6	5	1

Hint Look at the digits starting from the left.

Step 2   Compare the digits in the thousands place.

2 = 2

Step 3   Compare the digits in the hundreds place.

5 < 6

2,561 is less than 2,651.

Solution   2,561 < 2,651

Example 2

Order these numbers from greatest to least.

8,670   8,635   8,785

Step 1 Make a place-value chart to show the value of the digits in each number.

Thousands		Hundreds	Tens	Ones
8	,	6	7	0
8	,	6	3	5
8	,	7	8	5

Step 2 Compare the digits in the thousands place.

All of the numbers have 8 thousands.

Step 3 Compare the digits in the hundreds place.

6 < 7

8,785 is the greatest number.

Step 4 Compare the digits in the tens place.

7 > 3

8,670 is the next greatest number.

Hint Pay attention to what you are being asked to do. Are you ordering the numbers from least to greatest, or from greatest to least?

Solution The order of the numbers from greatest to least is 8,785, 8,670, 8,635.

Multiple-Choice Questions

1. Which symbol makes this number sentence true?

    3,492 _____ 3,618

    Test Taking Tip Cross out symbols that are not used to compare numbers.

    A. <   This symbol can be used to compare numbers. This could be right.
    B. >   This symbol can be used to compare numbers. This could be right.
    C. =   This symbol can be used to compare numbers. This could be right.
    D. +   This symbol cannot be used to compare numbers. Cross it out.

2. Which symbol makes this number sentence true?

    1,277 _____ 1,005

    A. <
    B. >
    C. =
    D. ×

3. Which symbol makes this number sentence true?

    4,389 _____ 3,796

    A. <
    B. >
    C. +
    D. =

4. Which symbol can be used to compare the two numbers below?

    7,240 _____ 7,240

    A. <
    B. >
    C. ×
    D. =

5. Which set of numbers is in order from greatest to least?

    A. 3,846; 3,589; 3,632
    B. 4,021; 4,360; 4,931
    C. 7,134; 7,116; 6,857
    D. 5,381; 5,421; 5,226

6. Which set of numbers is in order from least to greatest?

   A. 6,675; 6,438; 5,982
   B. 2,438; 2,701; 2,629
   C. 8,652; 8,431; 8,620
   D. 1,659; 1,671; 1,700

7. Which set of numbers is in order from least to greatest?

   A. 9,467; 9,468; 9,502
   B. 3,844; 3,823; 3,819
   C. 1,515; 1,849; 1,372
   D. 5,248; 4,913; 6,325

8. Which set of numbers is in order from greatest to least?

   A. 7,423; 730; 7,216
   B. 5,296; 5,237; 5,384
   C. 4,035; 4,392; 4,678
   D. 6,325; 6,249; 6,246

9. Which set of numbers is in order from greatest to least?

   A. 2,345; 2,861; 2,681
   B. 8,695; 8,324; 8,309
   C. 3,654; 3,982; 4,221
   D. 9,479; 9,801; 969

Short-Response Questions

10. Which symbol makes this number sentence true?

    1,834 _____ 1,843

11. A movie theatre sold the following number of tickets over three days:

    2,854   3,002   2,861

    Part A Write the numbers in order from least to greatest.

    _____

    Part B Write the numbers in order from greatest to least.

    _____

Lesson 12

The Associative Property of Multiplication

Math Words to Know

Associative Property of Multiplication a rule that states numbers grouped in any way will have the same product: $(7 \times 5) \times 2 = 7 \times (5 \times 2)$

Example 1

Multiply: $5 \times 2 \times 3$

> **Hint** If you can, regroup the numbers to make a simpler problem.

Step 1 Multiply 5×2 first.

$(5 \times 2) \times 3 = 10 \times 3$

Step 2 Multiply again.

$10 \times 3 = 30$

Solution $5 \times 2 \times 3 = 30$

Example 2

Multiply: (3 × 2) × 7

Hint Do the math in parentheses first.

Step 1 Multiply 3 × 2.

(3 × 2) × 7 = 6 × 7

Step 2 Multiply again.

6 × 7 = 42

Solution (3 × 2) × 7 = 42

Example 3

Write an expression that has the same product as (7 × 2) × 5.

Step 1 Write the expression without the parentheses.

7 × 2 × 5

Step 2 Put the parentheses around the last two factors.

7 × (2 × 5)

Hint The Associative Property of Multiplication states that the way factors are grouped does not change the product.

Solution 7 × (2 × 5) = (7 × 2) × 5

Multiple-Choice Questions

1. Multiply: 6 × (2 × 4)

   Test Taking Tip Estimate to cross out wrong answers.

   A. 12   This is 6 × 2. Cross it out.
   B. 14   This answer is too low. Cross it out.
   C. 36   This is 6 × 6. This could be right.
   D. 48   This is 6 × 8. This could be right.

2. Multiply: (5 × 4) × 3
   A. 12
   B. 35
   C. 60
   D. 80

3. Multiply: 3 × 3 × 5
   A. 9
   B. 15
   C. 45
   D. 60

4. Multiply: 2 × (5 × 6)
   A. 10
   B. 30
   C. 45
   D. 60

5. Which expression has the same product as 9 × (7 × 5)?
   A. (9 × 7) × 5
   B. 9 + (7 + 5)
   C. 9 + (7 × 5)
   D. (9 × 7) + 5

6. Which expression has the same product as (3 × 2) × 9?

   A. 3 × (2 × 9)

   B. (3 + 2) + 9

   C. (3 + 2 × 9)

   D. (3 + 2) × 9

7. Which expression has the same product as 7 × (4 × 8)?

   A. 7 × (4 + 8)

   B. (7 × 4) × 8

   C. 7 + (4 × 8)

   D. (7 + 4) × 8

8. Which expression has the same product as (1 × 6) × 5?

   A. 0 × (6 × 5)

   B. (1 + 6) + 5

   C. 1 × (6 × 5)

   D. (1 + 6) × 5

9. Which expression has the same product as 2 × (9 × 8)?

   A. 2 × (9 + 8)

   B. 2 + (9 + 8)

   C. (2 × 9) × 8

   D. (2 + 9) × 8

Short-Response Questions

10. Multiply: 9 × (2 × 5)

    Answer _____

11. **Part A**   Multiply: (6 × 2) × 4

    Answer _____

    Part B   Write the expression another way and multiply to check your answer to Part A.

    Answer _____

Lesson 13

Properties of Multiplying Even and Odd Numbers

Math Words to Know

even number a number that divides evenly by 2; the ones digit of an even number is always 0, 2, 4, 6, or 8

odd number a number that does not divide evenly by 2; the ones digit of an odd number is always 1, 3, 5, 7, or 9

Example 1

Is the product of 4 × 6 even or odd?

Step 1 Multiply.

4 × 6 = 24

Step 2 Decide if 24 is even or odd. Look at the ones digit.

The ones digit is 4. Four is even.

Hint The product of two even numbers is always even.

Solution The product of 4 × 6 is even.

Example 2

Is the product of 3 × 9 even or odd?

Step 1 Multiply.

3 × 9 = 27

Step 2 Decide if 27 is even or odd. Look at the ones digit.

The ones digit is 7. Seven is odd.

Hint The product of two odd numbers is always odd.

Solution The product of 3 × 9 is odd.

Example 3

Which number could you multiply by 5 to get an even product?

3   8

Step 1 Multiply.

5 × 3 = 15

15 is odd.

Step 2 Multiply.

5 × 8 = 40

40 is even.

Hint The product of one even number and one odd number is always even.

Solution I could multiply 5 × 8 to get an even product.

Multiple-Choice Questions

1. Choose the pair of factors that has an odd product.

   Test Taking Tip Find the answer with two odd factors.

A.	7 × 8	One factor is even. Cross it out.
B.	3 × 6	One factor is even. Cross it out.
C.	4 × 2	Both factors are even. Cross it out.
D.	5 × 3	Both factors are odd. This could be the right answer.

2. Choose the pair of factors that has an even product.

   A. 9 × 1
   B. 6 × 4
   C. 3 × 3
   D. 7 × 5

3. Choose the pair of factors that has an odd product.

   A. 7 × 3
   B. 2 × 5
   C. 8 × 4
   D. 1 × 6

4. Choose the pair of factors that has an even product.

   A. 7 × 7
   B. 9 × 3
   C. 5 × 6
   D. 1 × 5

5. Which number could you multiply by 7 to get an odd product?

   A. 0
   B. 5
   C. 6
   D. 8

58   Number Sense and Operations

6. Which number could you multiply by 3 to get an even product?

   A. 2
   B. 3
   C. 5
   D. 9

7. Which number could you multiply by 5 to get an odd product?

   A. 2
   B. 5
   C. 6
   D. 8

8. Choose the pair of factors that has an even product.

   A. 3×5
   B. 7×9
   C. 1×1
   D. 6×2

9. Which number could you multiply by 9 to get an even product?

   A. 1
   B. 3
   C. 5
   D. 6

Short-Response Questions

10. Is the product of 7×6 even or odd?

    Answer _____

11. **Part A**   Is the product of 8×8 even or odd?

    Answer _____

    Part B   Tell how you know that the product in Part A is odd or even without multiplying.

    _____
    _____
    _____

Lesson 13: Properties of Multiplying Even and Odd Numbers   59

Lesson 14

Adding and Subtracting Whole Numbers

Math Words to Know

addition combining groups to find the total amount

subtraction taking away to find a difference

whole number a number that is not a fraction or decimal (0, 1, 2, 3, etc.)

counting number 1, 2, 3, etc.

Example 1

64 + 37 = _____

Step 1 Rewrite the problem. Line up the digits by their place value.

```
   64
+  37
```

Step 2 Add the ones.

Hint Regroup when the sum is 10 or greater.

```
   1
   64
+  37
─────
    1
```

$4 + 7 = 11$

Regroup 11 as 1 ten and 1 one.

Step 3 Add the tens.

```
   1
   64
+  37
─────
  101
```

$1 + 6 + 3 = 10$

Solution 64 + 37 = 101

Example 2

87 − 52 = _____

Step 1 Rewrite the problem. Line up the digits by their place value.

$$\begin{array}{r} 87 \\ -52 \\ \hline \end{array}$$

Step 2 Subtract the ones.

$$\begin{array}{r} 87 \\ -52 \\ \hline 5 \end{array}$$ 7 − 2 = 5

Step 3 Subtract the tens.

$$\begin{array}{r} 87 \\ -52 \\ \hline 35 \end{array}$$ 8 − 5 = 3

Hint You can check your subtraction answer by adding. Does 35 + 52 = 87?

Solution 87 − 52 = 35

Lesson 14: Adding and Subtracting Whole Numbers

Multiple-Choice Questions

1. 46 + 32 = _____

Test Taking Tip Solve a simpler problem. Start by adding the ones digits.

- **A.** 78   6 + 2 = 8. This could be right.
- **B.** 79   6 + 2 does not equal 9. Cross it out.
- **C.** 88   6 + 2 = 8. This could be right.
- **D.** 89   6 + 2 does not equal 9. Cross it out.

2. 57 + 21 = _____
- **A.** 36
- **B.** 78
- **C.** 79
- **D.** 88

3. 39 + 43 = _____
- **A.** 72
- **B.** 81
- **C.** 82
- **D.** 92

4. 67 + 18 = _____
- **A.** 49
- **B.** 72
- **C.** 75
- **D.** 85

5. 75 − 35 = _____
- **A.** 40
- **B.** 45
- **C.** 50
- **D.** 110

6. 98 − 56 = _____
   A. 32
   B. 38
   C. 41
   D. 42

7. 42 − 17 = _____
   A. 25
   B. 35
   C. 37
   D. 59

8. 60 − 23 = _____
   A. 27
   B. 33
   C. 35
   D. 37

9. 84 − 69 = _____
   A. 5
   B. 15
   C. 35
   D. 36

Short-Response Questions

10. 77 − 38 = _____

11. 54 + 19 = _____

Lesson 15

Models for Multiplication and Division

Math Words to Know

row a line in a table or array that goes across

column a line in a table or array that goes up and down

Example 1

Which multiplication sentence does this model show?

Step 1 Write a blank multiplication sentence.

_____ × _____ = _____

Hint The number of rows and columns are factors in the multiplication sentence.

Step 2 Count the number of rows.

There are 2 rows.

2 × _____ = _____

Step 3 Count the number of columns.

There are 7 columns.

2 × 7 = _____

Step 4 Count the number of total objects.

There are 14 books.

2 × 7 = **14**

Solution The model shows the multiplication sentence 2 × 7 = 14.

Example 2

Draw a model to show 18 ÷ 6.

Step 1 Divide to find the quotient.

18 ÷ 6 = 3

Step 2 The dividend shows the total number of objects.

There are 18 total objects.

Step 3 The divisor can be used to show the number of rows.

There are 6 rows.

Step 4 The quotient can be used to show the number of columns.

There are 3 columns.

Solution This model shows 18 ÷ 6:

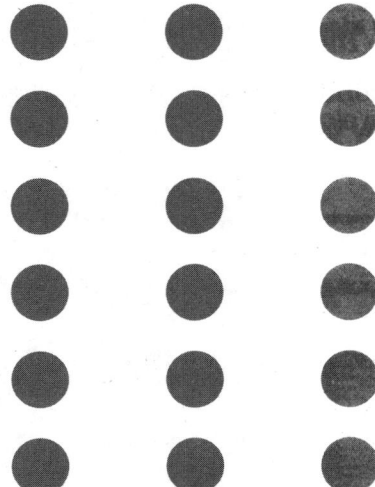

Multiple-Choice Questions

1. Which expression does the model show?

A.	4 = 5	The model does not show a comparison. Cross it out.
B.	4 × 5	The model can be used to show a multiplication expression. This could be right.
C.	20 − 4	The model does not show a subtraction expression. Cross it out.
D.	5 ÷ 4	The model can be used to show a division expression. This could be right.

Test Taking Tip Does this model show +, −, ×, or ÷?

2. Which expression does the model show?

A. 3 × 9
B. 3 + 9
C. 9 ÷ 3
D. 27 − 9

3. Which expression does the model show?

A. 40 − 5
B. 48 ÷ 8
C. 6 × 8
D. 40 ÷ 5

4. Which model shows 4 × 7?

   A. (pencils arranged)

   B. (pencils arranged)

   C. (pencils arranged)

   D. (pencils arranged)

5. Which model shows 30 ÷ 6?

   A. (cats arranged 5 × 5)

   B. (cats arranged 5 × 6)

   C. (cats arranged)

   D. (cats arranged)

Short-Response Question

6. **Part A** Draw a model to show 2 × 9.

   Part B How many objects did you draw?

   Answer _____

Lesson 15: Models for Multiplication and Division 67

Lesson 16

Relating Multiplication and Division

Math Words to Know

multiplication fact   a multiplication sentence, such as $2 \times 4 = 8$

division fact   a division sentence, such as $8 \div 2 = 4$

fact family   facts that are related, using the same numbers

Example 1

Write a multiplication fact that is related to $9 \times 4 = 36$.

Step 1   What are the factors?

The factors are 9 and 4.

Step 2   Switch the order of the factors.

4×9

Hint   The order of the factors does not change the product in a multiplication fact.

Solution   The multiplication fact $4 \times 9 = 36$ is related to $9 \times 4 = 36$.

Example 2

Write a division fact that is related to 42 ÷ 6 = 7.

Step 1 Which are the divisor and quotient?

The divisor is 6 and the quotient is 7.

Step 2 Switch the divisor and quotient.

42 ÷ 7 = 6

Solution The division fact 42 ÷ 7 = 6 is related to 42 ÷ 6 = 7.

Example 3

Write two multiplication facts using the numbers 8, 24, and 3.

> **Hint** The greatest number is always the product in a multiplication fact.

Step 1 Write 24 as the product for two multiplication sentences.

_____ × _____ = 24

_____ × _____ = 24

Step 2 Insert the factors into the first multiplication fact.

8 × 3 = 24

Step 3 Switch the order of the factors. Put them into the second multiplication fact.

8 × 3 = 24

3 × 8 = 24

Solution Two multiplication facts using the numbers 8, 24, and 3, are
8 × 3 = 24 and 3 × 8 = 24.

Multiple-Choice Questions

1. Which multiplication fact is related to $3 \times 5 = 15$?

   Test Taking Tip What are you asked to find?
   Cross out answers that are not multiplication facts.

   A. $3 \times 6 = 18$ — This is a multiplication fact. This could be right.

   B. $15 \div 5 = 3$ — This is not a multiplication fact. Cross it out.

   C. $5 + 5 + 5 = 15$ — This is not a multiplication fact. Cross it out.

   D. $5 \times 3 = 15$ — This is a multiplication fact. This could be right.

2. Which multiplication fact is related to $10 \times 2 = 20$?

   A. $2 \times 10 = 20$

   B. $10 + 10 = 20$

   C. $2 \times 11 = 22$

   D. $20 \div 10 = 2$

3. Which multiplication sentence is related to $6 \times 4 = 24$?

   A. $8 \times 3 = 24$

   B. $4 \times 6 = 24$

   C. $6 \times 5 = 30$

   D. $6 + 6 + 6 + 6 = 24$

4. Which division fact is related to $56 \div 7 = 8$?

   A. $8 \div 7 = 56$

   B. $56 \div 56 = 1$

   C. $56 \div 8 = 7$

   D. $56 - 8 - 8 - 8 - 8 - 8 - 8 - 8 = 0$

5. Which division sentence is related to $12 \div 6 = 2$?

   A. $2 \times 6 = 12$

   B. $12 \div 4 = 3$

   C. $12 - 6 - 6 = 0$

   D. $12 \div 2 = 6$

6. Which division fact is related to 45 ÷ 9 = 5?

   A. 54 ÷ 6 = 9
   B. 45 ÷ 5 = 9
   C. 45 − 9 − 9 − 9 − 9 − 9 = 0
   D. 9 ÷ 5 = 45

7. Which multiplication fact can be written using the numbers 28, 7, and 4?

   A. 7 × 4 = 28
   B. 4 × 28 = 7
   C. 28 ÷ 7 = 4
   D. 7 + 7 + 7 + 7 = 28

8. Which multiplication fact can be written using the numbers 8, 80, and 10?

   A. 10 + 10 + 10 + 10 + 10 + 10 + 10 + 10 = 80
   B. 8 × 80 = 10
   C. 10 × 8 = 80
   D. 80 ÷ 8 = 10

9. Which multiplication fact can be written using the numbers 4, 3, and 12?

   A. 6 × 2 = 12
   B. 4 + 4 + 4 = 12
   C. 12 ÷ 4 = 3
   D. 3 × 4 = 12

Short-Response Questions

10. Write a division fact that is related to 20 ÷ 4 = 5.

    Answer _____

11. **Part A** Write two multiplication facts using the numbers 5, 35, and 7.

    Answer _____

    Part B Write two division facts using the numbers 5, 35, and 7.

    Answer _____

Lesson 16: Relating Multiplication and Division   71

Lesson 17

Multiplying and Dividing by Multiples of 10 and 100

Math Words to Know

multiple   the product of a number and any whole number

Example 1

What is 400 × 10?

> Step 1 How many zeros are in 10?
>
> There is one zero in 10.
>
> **Hint** When multiplying a number by 10, put 1 zero after the number to get the product.
>
> Step 2 Insert the zero to the right of the first factor.
>
> 400**0**

Solution   400 × 10 = 4,000

Example 2

Multiply: 300 × 50

Step 1 Multiply the non-zero digits.

$3 \times 5 = 15$

Step 2 Count the number of zeros in the factors.

300 has 2 zeros. 50 has 1 zero.

There are 3 zeros in all.

Step 3 Put the 3 zeros after the product (15).

15000

Solution $300 \times 50 = 15{,}000$

Example 3

Divide: 800 ÷ 40

Step 1 How many zeros are in the divisor?

There is 1 zero in 40.

Step 2 Take away the zero from **both** the dividend and divisor.

800 ⟶ 80

40 ⟶ 4

Hint When a dividend and divisor both have zeros, take away the same number of zeros from both numbers.

Step 3 Divide the numbers.

$80 \div 4 = 20$

Hint You can multiply the quotient and the divisor to check your answer. $20 \times 4 = 80$.

Solution $800 \div 40 = 20$

Multiple-Choice Questions

1. Multiply: 600 × 10

   Test Taking Tip Count the number of zeros in both factors.

   A. 60   This answer has 1 zero. Cross it out.
   B. 6,000   This answer has 3 zeros. This could be right.
   C. 7,000   This answer has 3 zeros. This could be right.
   D. 60,000   This answer has 4 zeros. Cross it out.

2. Multiply: 200 × 10
   A. 20
   B. 2,000
   C. 3,000
   D. 20,000

3. Multiply: 300 × 10
   A. 30
   B. 300
   C. 3,000
   D. 4,000

4. Multiply: 700 × 40
   A. 2,800
   B. 11,000
   C. 28,000
   D. 280,000

5. Multiply: 500 × 80
   A. 1,300
   B. 4,000
   C. 13,000
   D. 40,000

6. Multiply: 900 × 60

    A.   5,400
    B.   15,000
    C.   54,000
    D.   540,000

7. Divide: 200 ÷ 10

    A.   2
    B.   20
    C.   200
    D.   2,000

8. Divide: 900 ÷ 30

    A.   30
    B.   300
    C.   3,000
    D.   27,000

9. Divide: 800 ÷ 20

    A.   4
    B.   8
    C.   20
    D.   40

Short-Response Questions

10. Divide: 700 ÷ 70

    Answer _____

11. **Part A**   Multiply: 500 × 40

    Answer _____

    Part B   Multiply: 500 × 400

    Answer _____

Lesson 18

Multiplying Two-Digit Numbers by One-Digit Whole Numbers

Math Words to Know

product   the answer in a multiplication problem

Example 1

Multiply: 73 × 2

Step 1   Rewrite the problem. Line up the digits by their place value.

Hint Write the two-digit number on top and the one-digit number on the bottom.

$$\begin{array}{r} 73 \\ \times\ 2 \\ \hline \end{array}$$

Step 2   Multiply 2 by the ones digit.

$$\begin{array}{r} 73 \\ \times\ 2 \\ \hline 6 \end{array} \qquad 2 \times 3 = \mathbf{6}$$

Step 3   Multiply 2 by the tens digit.

$$\begin{array}{r} 73 \\ \times\ 2 \\ \hline \mathbf{14}6 \end{array} \qquad 2 \times 7 = \mathbf{14}$$

Solution   73 × 2 = 146

Example 2

Find the product of 51 and 9.

> **Hint** "Find the product" means to multiply.

Step 1 Rewrite the problem. Line up the digits by their place value.

$$\begin{array}{r} 51 \\ \times9 \\ \hline \end{array}$$

Step 2 Multiply 9 by the ones digit.

$$\begin{array}{r} 51 \\ \times9 \\ \hline 9 \end{array} \qquad 9 \times 1 = \mathbf{9}$$

Step 3 Multiply 9 by the tens digit.

$$\begin{array}{r} 51 \\ \times9 \\ \hline \mathbf{45}9 \end{array} \qquad 9 \times 5 = \mathbf{45}$$

Solution The product of 51 and 9 is 459.

Multiple-Choice Questions

1. Multiply: 64 × 2

   Test Taking Tip Estimate by multiplying 60 × 2 in your head. Cross out answers that do not make sense.

   A. 66   This answer is too low. Cross it out.

   B. 88   This answer is too low. Cross it out.

   C. 126   This answer is close to 60 × 2. This could be right.

   D. 128   This answer is close to 60 × 2. This could be right.

2. Multiply: 53 × 3

   A. 56

   B. 156

   C. 159

   D. 189

3. Multiply: 81 × 6

   A. 86

   B. 486

   C. 487

   D. 496

4. Multiply: 94 × 2

   A. 188

   B. 194

   C. 196

   D. 198

5. Multiply: 42 × 4

   A. 46

   B. 166

   C. 168

   D. 188

6. What is the product of 72 and 3?

    A. 75
    B. 146
    C. 215
    D. 216

7. What is the product of 61 and 8?

    A. 69
    B. 481
    C. 488
    D. 489

8. What is the product of 24 and 2?

    A. 48
    B. 84
    C. 86
    D. 88

9. What is the product of 31 and 7?

    A. 37
    B. 38
    C. 217
    D. 218

Short-Response Questions

10. Multiply: 43×3

    Answer _____

11. What is the product of 83 and 2?

    Answer _____

Lesson 19

Dividing Two-Digit Dividends by One-Digit Divisors

Math Words to Know

dividend in division, the number that is divided

divisor in division, the number that you divide by

quotient in division, the answer to the problem

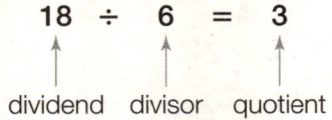
dividend divisor quotient

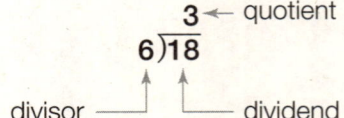
divisor dividend quotient

Example 1

Divide: 46 ÷ 2

Step 1 Divide the tens, then multiply.

$$\begin{array}{r} 2 \\ 2\overline{)46} \\ 4 \end{array}$$ ⟶ 2 × 2 = 4

Step 2 Subtract and bring down the second number of the dividend.

$$\begin{array}{r} 2 \\ 2\overline{)46} \\ -4 \\ \hline 06 \end{array}$$ ⟶ 4 − 4 = 0

Step 3 Divide the ones, then multiply and subtract.

$$\begin{array}{r} 23 \\ 2\overline{)46} \\ -4 \\ \hline 06 \\ -6 \\ \hline 0 \end{array}$$
06 ⟶ 3 × 2 = 6
−6 ⟶ 6 − 6 = 0

Solution 46 ÷ 2 = 23

Example 2

A package of hamburger buns has 8 buns. Jorge is making 32 hamburgers. How many packages of buns does Jorge need?

Step 1 Write the problem.

Jorge is making 32 hamburgers. There are 8 buns in a package.

The problem is $32 \div 8$.

Step 2 Divide.

Divide 3 by 8. Since 8 is greater than 3, you cannot divide at this point.

$$8\overline{)32}$$

Hint When the divisor is greater than the first number in the dividend, use the first **two** numbers in the dividend.

Step 3 Divide 32 by 8.

$$8\overline{)32}^{\,4} \qquad 32 \div 8 = 4$$

Solution Jorge needs 4 packages of hamburger buns.

Multiple-Choice Questions

1. Divide: 96 ÷ 3

   Test Taking Tip Estimate by dividing 90 ÷ 3. Cross out the answers that are not close.

   A. 32   This is close to 90 ÷ 3. This could be right.

   B. 33   This is close to 90 ÷ 3. This could be right.

   C. 63   This answer does not make sense. Cross it out.

   D. 93   This is 96 − 3. Cross it out.

2. Divide: 48 ÷ 4

   A. 11

   B. 12

   C. 22

   D. 44

3. Divide: 50 ÷ 5

   A. 1

   B. 2

   C. 5

   D. 10

4. Divide: 86 ÷ 2

   A. 43

   B. 44

   C. 63

   D. 84

5. Divide: 62 ÷ 2

   A. 30

   B. 31

   C. 32

   D. 60

82 Number Sense and Operations

6. Divide: 56 ÷ 7

    A. 6
    B. 8
    C. 9
    D. 49

7. Divide: 35 ÷ 5

    A. 6
    B. 7
    C. 15
    D. 30

8. There are 28 word problems. 4 students in a group divide the word problems evenly. Divide 28 ÷ 4 to find how many word problems each student does.

    A. 4 word problems
    B. 6 word problems
    C. 7 word problems
    D. 24 word problems

9. Nadia has 72 stickers in 9 packages. How many stickers are in each package? Divide 72 ÷ 9.

    A. 5 stickers
    B. 6 stickers
    C. 7 stickers
    D. 8 stickers

Short-Response Questions

10. Divide: 93 ÷ 3

    Answer _____

11. David wants to put 42 songs onto 6 CDs. He wants the same number of songs on each CD. How many songs can David put on each CD? Divide 42 ÷ 6.

    Answer _____ songs

Lesson 20

How to Solve Problems

Math Words to Know

operation addition, subtraction, multiplication, or division

Example 1

Diana typed 76 words in one minute. Jordan typed 68 words in one minute. How do you find how many more words Diana typed than Jordan?

Step 1 Which operation will you use to solve the word problem?

Hint Look for key words such as **how many more**, **how much more**, and **how many are left**. These words mean that you will subtract.

The words **how many more** means you will subtract.

Step 2 Which numbers will you subtract?

76 and 68

Step 3 How will you subtract?

I will subtract 68 **from** 76.

Hint When subtracting, you always subtract the smaller number **from** the larger number.

Solution Subtract 68 from 76 to find how many more words Diana typed than Jordan.

84 Number Sense and Operations

Example 2

Ms. Lu's class gave 89 cans of food to charity. Mr. Parra's class gave 65 cans of food. How many cans of food did both classes give in all?

Step 1   Which operation will you use to solve the word problem?

> **Hint** Look for key words such as **total**, **in all**, and **how many**. These words mean that you will add.

The words **in all** means you will add.

Step 2   Write the addition problem.

```
   89
+  65
-----
```

Step 3   Add.

```
   1
   89
+  65
-----
  154
```

Solution   The classes gave 154 cans in all.

Lesson 20: How to Solve Problems   85

Multiple-Choice Questions

1. Olivia picked 36 apples at a farm. Javier picked 51 apples. How do you find how many more apples Javier picked than Olivia?

   A. Add 36 and 51.
   B. Subtract 36 from 51.
   C. Multiply 36 by 51.
   D. Subtract 51 from 36.

   Test Taking Tip Which operation should you use?
   You do not need to add. Cross it out.
   You need to subtract. This could be right.
   You do not need to multiply. Cross it out.
   You need to subtract. This could be right.

2. Sarah read one book for 12 hours on vacation. She read another book for 13 hours. How do you find how many total hours Sarah read on vacation?

   A. Add 12 and 13.
   B. Subtract 12 from 13.
   C. Subtract 13 from 12.
   D. Multiply 12 by 13.

3. Brandon got 76 right answers on his test. Anna got 84 right answers. How do you find how many more right answers Anna had than Brandon?

   A. Divide 84 by 76.
   B. Add 76 and 84.
   C. Subtract 84 from 76.
   D. Subtract 76 from 84.

4. Greg has 39 marbles. He bought 52 more. How many total marbles does Greg have now?

   A. 13 marbles
   B. 81 marbles
   C. 91 marbles
   D. 93 marbles

5. Julie and her mom baked 45 granola bars. Her family ate 27 bars. How many granola bars are left?

   A. 18 bars
   B. 28 bars
   C. 29 bars
   D. 72 bars

6. Hector has 70 stamps in his album. He gave his sister 34 stamps. How many stamps does Hector have now?

   A. 26 stamps

   B. 36 stamps

   C. 46 stamps

   D. 106 stamps

7. Mr. Chu's class raised $69 for charity. Ms. Hernandez's class raised $58. How much did the classes raise in all?

   A. $11

   B. $117

   C. $125

   D. $127

8. Laura caught a fish that weighed 14 pounds. She caught another fish that weighed 26 pounds. How much did the two fish Laura caught weigh in all?

   A. 12 pounds

   B. 40 pounds

   C. 42 pounds

   D. 50 pounds

9. Jaime and his mom sold 45 scarves at an arts and crafts sale. They also sold 82 blankets. How many more blankets than scarves did they sell?

   A. 37 blankets

   B. 46 blankets

   C. 47 blankets

   D. 127 blankets

Short-Response Questions

10. Tanya was given $35 for her birthday. She spent $27 on a jacket.

    Part A How would you find how much money Tanya has left?

    Answer _____

    Part B How much money does she have left?

    Answer $ _____

11. Ryan and his family drove 91 miles to see his grandmother. They drove another 63 miles to visit his aunt. How many total miles did Ryan and his family drive?

    Answer _____ miles

Lesson 21

Rounding to the Nearest Ten and Hundred

Math Words to Know

rounding to estimate the value of a number based on a given place value

Rounding to the nearest ten, 2<u>4</u>8 rounds up to 250.

Rounding to the nearest hundred, <u>2</u>48 rounds down to 200.

Example 1

Round 54 to the nearest ten.

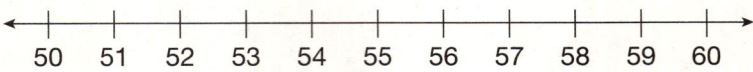

Hint You can use number lines to round numbers.

Step 1 Mark 54 on the number line.

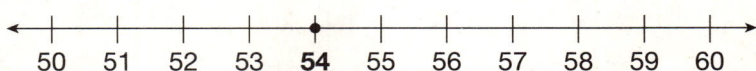

Step 2 Is 54 closer to 50 or 60 on the number line?

Hint If the ones number is less than 5, round down. If the ones number is 5 or greater, round up.

54 is closer to 50 than 60.

Solution 54 rounded to the nearest 10 is 50.

88 Number Sense and Operations

Example 2

Round 286 to the nearest hundred.

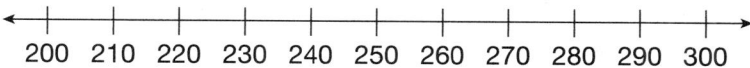

Step 1 Mark 286 on the number line.

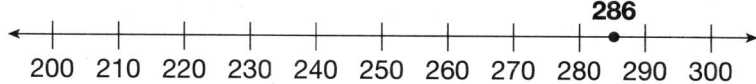

Step 2 Is 286 closer to 200 or 300 on the number line?

286 is closer to 300 than 200.

Solution 286 rounded to the nearest hundred is 300.

Example 3

Round 25 to the nearest ten.

Step 1 Mark 25 on the number line.

Step 2 Is 25 closer to 20 or 30 on the number line?

25 is exactly halfway between 20 and 30.

Step 3 Round 25 up to the nearest 10.

Hint Anytime a number is exactly halfway between two numbers, round **up**.

Solution 25 rounded to the nearest ten is 30.

Lesson 21: Rounding to the Nearest Ten and Hundred

Multiple-Choice Questions

1. Round 71 to the nearest ten.

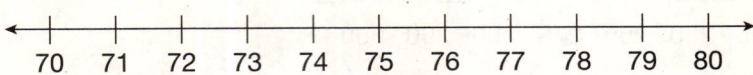

Test Taking Tip Mark 71 on the number line. Which ten is it closest to?

A. 60   71 is not near 60. Cross it out.
B. 70   71 is very close to 70. This could be the right answer.
C. 80   71 is not near 80. Cross it out.
D. 100 71 is not near 100. Cross it out.

2. Round 38 to the nearest ten.

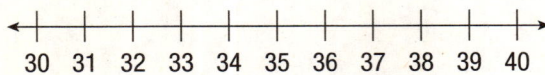

A. 30
B. 35
C. 40
D. 100

3. Round 65 to the nearest ten.

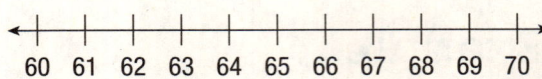

A. 50
B. 60
C. 65
D. 70

90   Number Sense and Operations

4. Round 542 to the nearest hundred.

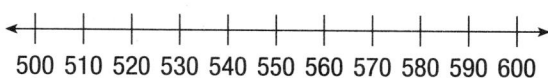

- A. 500
- B. 540
- C. 550
- D. 600

5. Round 857 to the nearest hundred.

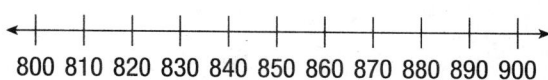

- A. 800
- B. 850
- C. 860
- D. 900

Short-Response Questions

6. Miguel drew a number line from 100 to 200 on the board and made a mark at 150.

   Part A Draw the number line with a mark at 150.

   Part B Round 150 to the nearest hundred.

   Answer _____

7. Round 63 to the nearest ten.

   Answer _____

Lesson 22

Using Estimation to Check Answers

Math Words to Know

estimate a number that is close to the exact amount

Example 1

The Drama Club sold 384 tickets to their first play. They sold 529 tickets to their second play. How would you find an estimate of how many total tickets the club sold?

Step 1 Which operation will you use to solve the word problem?

This problem can be solved using addition.

Step 2 Round both numbers to the nearest 100.

Hint Use number lines to help you round.

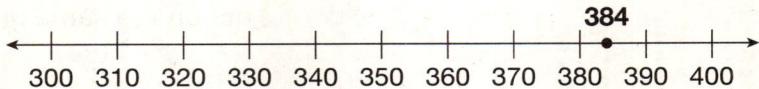

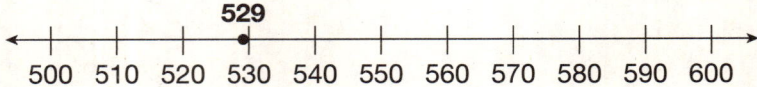

384 rounds to 400.

529 rounds to 500.

Solution You can find an estimate by adding 400 + 500.

Example 2

Stephanie has 78 seashells. Robert has 32 seashells. Estimate how many more seashells Stephanie has than Robert.

Step 1 Which operation will you use to solve the word problem?

This problem can be solved using subtraction.

Step 2 Round both numbers to the nearest 10.

78 rounds to 80.

32 rounds to 30.

Step 3 Write the subtraction problem. Subtract.

80 − 30 = 50

Solution Stephanie has about 50 more seashells than Robert.

Example 3

Chris scored 780 points on a video game. Bianca scored 418 points on the same game. Chris said he scored 362 more points than Bianca. Is Chris's answer reasonable?

Step 1 Which operation will you use to solve the word problem?

This problem can be solved using subtraction.

Step 2 Round both numbers to the nearest 100.

780 rounds to 800.

418 rounds to 400.

Hint An answer is reasonable if the estimate is close to the answer.

Step 3 Write the subtraction problem. Subtract.

800 − 400 = 400

Solution Chris's answer is close to the estimate. His answer is reasonable.

Lesson 22: Using Estimation to Check Answers

Multiple-Choice Questions

1. Terry has $97 in her savings account. She also has $69 in her piggy bank. How would you find an estimate of how much more money Terry has in her savings account than in her piggy bank?

   A. Add 100 and 70.

   B. Subtract 70 from 100.

   C. Divide 100 by 70.

   D. Subtract 70 from 90.

   Test Taking Tip Which operation should you use?
   You do not need to add. Cross it out.
   You need to subtract. This could be right.
   You do not need to divide. Cross it out.
   You need to subtract. This could be right.

2. On Monday, 227 people came to a town meeting. On Tuesday, 318 people came to the meeting. How would you find an estimate of how many people came to the meetings all together?

   A. Add 200 and 300.

   B. Subtract 200 from 300.

   C. Add 300 and 300.

   D. Add 200 and 200.

3. In the school election, 65 students voted for Jessica for class president and 43 students voted for Brad. How would you find an estimate of how many more people voted for Jessica than for Brad?

   A. Subtract 40 from 60.

   B. Add 60 and 40.

   C. Add 70 and 40.

   D. Subtract 40 from 70.

4. Jennifer planted 158 vegetable seeds in her garden. She also planted 347 flower seeds in the garden. Estimate how many total seeds Jennifer planted in her garden.

   A. 400 seeds

   B. 450 seeds

   C. 500 seeds

   D. 600 seeds

5. Ramon read a book that had 294 pages. He read another book that had 207 pages. Estimate how many total pages Ramon read.

   A. 300 pages

   B. 400 pages

   C. 500 pages

   D. 600 pages

6. One year, Buffalo, NY, got 94 inches of snow. That same year, New York City got 27 inches of snow. Estimate how many more inches of snow Buffalo got than New York City.

   A. 60 inches
   B. 80 inches
   C. 110 inches
   D. 120 inches

7. Ms. Lopez rode 835 miles on her bicycle in June. She rode her bicycle 649 miles in July. Ms. Lopez said she rode her bike 186 more miles in June than in July. Is her answer reasonable?

   A. Yes, because $800 - 700 = 100$.
   B. No, because $800 + 600 = 1,400$.
   C. No, because $900 - 600 = 300$.
   D. Yes, because $800 - 600 = 200$.

8. Mr. Wang bought 74 blank CDs last week. He bought 68 blank CDs this week. Mr. Wang said he bought 142 blank CDs for the two weeks. Is his answer reasonable?

   A. No, because $80 - 70 = 10$.
   B. Yes, because $70 + 70 = 140$.
   C. No, because $70 + 60 = 130$.
   D. Yes, because $80 + 70 = 150$.

9. On Thursday, 426 people saw a baseball game. On Friday, 509 people saw another game. The newspaper said 1,000 total people saw the two games. Is the newspaper's estimate right?

   A. No, because $400 + 500 = 900$.
   B. Yes, because $500 + 500 = 1,000$.
   C. No, because $500 - 400 = 100$.
   D. Yes, because $400 + 500 = 1,000$.

Short-Response Question

10. Patrick cut a length of string that measured 150 centimeters. He cut another piece that measured 408 centimeters. Patrick said he cut 658 centimeters of string. Is his answer reasonable? Use estimation. Tell why or why not.

_____

_____

_____

Lesson 22: Using Estimation to Check Answers

PROGRESS CHECK 1

1 Which shows 32 ÷ 4?

A 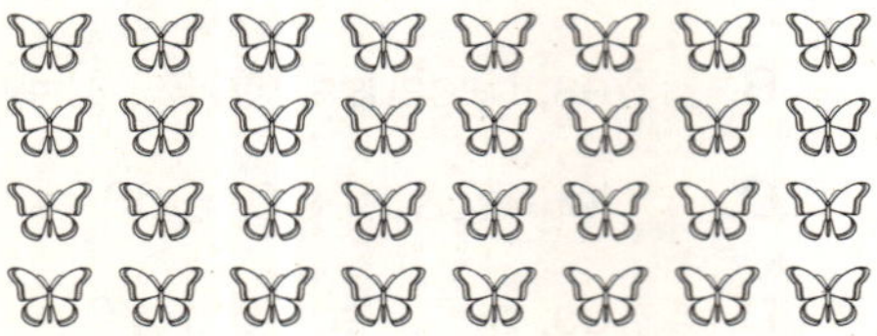

B 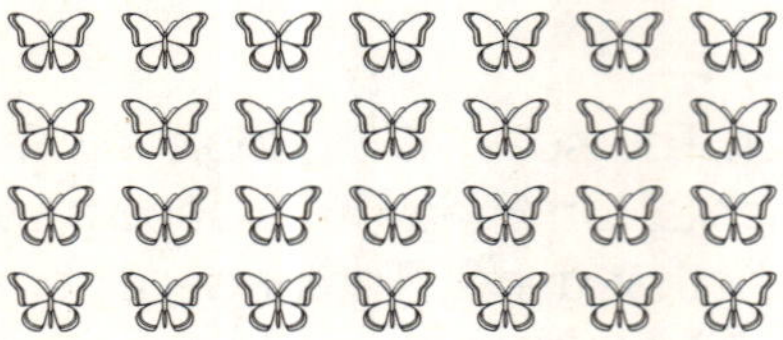

C 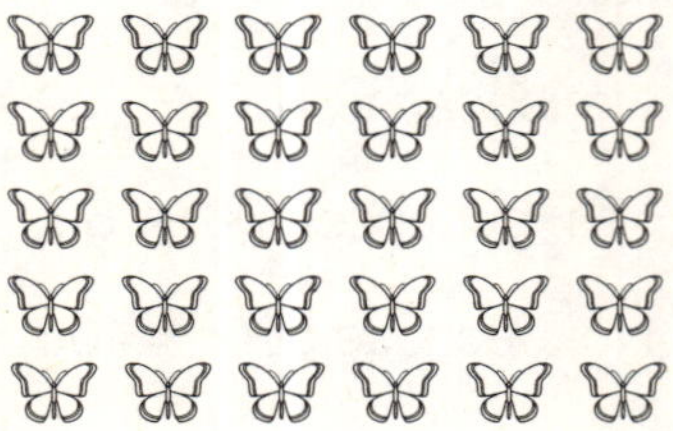

D 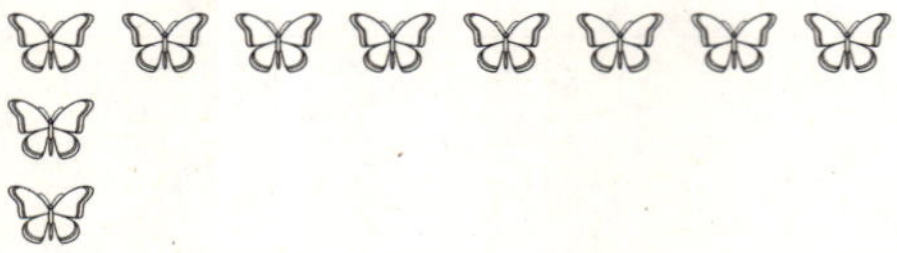

2 Round 653 to the nearest hundred.

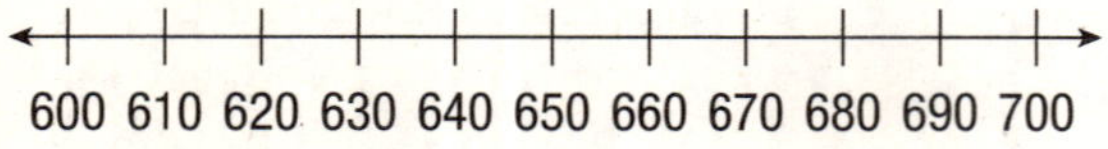

F 600

G 650

H 660

J 700

3 What is the value of 8 in 8,036?

A 8

B 800

C 8,000

D 80,000

4 Multiply: 82 × 3

F 85

G 245

H 246

J 326

5 Which expression has the same product as 6 × (4 × 8)?

A (6 × 4) × 8

B 6 + (4 + 8)

C 6 + (4 × 8)

D (6 × 4) + 8

6 Which pair of factors has an even product?

  F 7 × 1   H 7 × 7
  G 3 × 5   J 9 × 4

7 Which division sentence is related to 36 ÷ 9 = 4?

  A 9 × 4 = 36
  B 36 ÷ 36 = 1
  C 36 − 9 − 9 − 9 − 9 = 0
  D 36 ÷ 4 = 9

8 Multiply: 500 × 90

  F 1,400   H 14,000
  G 4,500   J 45,000

9 Which symbol makes this sentence true?

  9,367 _____ 9,376

  A <   C =
  B >   D +

Short-Response Question

10 Natasha put 88 raisins in her trail mix. She put 43 peanuts in the trail mix.

Part A

How would you find how many more raisins than peanuts are in the trail mix?

_____

_____

Part B

How many more raisins than peanuts are in the trail mix?

Show your work.

Answer _____ raisins

PROGRESS CHECK 2

1 Which of the following numbers could you multiply by 3 to have an odd product?

A 2

B 6

C 7

D 8

2 Which set of numbers is in order from least to greatest?

F 6,254; 7,103; 6,836

G 1,595; 1,395; 1,359

H 4,078; 4,089; 4,108

J 2,426; 2,311; 2,375

3 A radio station took 35 calls one hour. The next hour, the radio station took 49 calls. How do you find the total number of calls the radio station took for the two hours?

A Multiply 2 by 35.

B Add 35 and 49.

C Subtract 35 from 49.

D Multiply 35 by 49.

4 A beach had 5,348 swimmers one weekend. What is the word form of this number?

F five thousand, three forty-eight

G five thousand, three hundred forty-eight

H five thousand, three hundred fourty-eighty

J fifty-three thousand forty-eight

5 6,572 − 3,427 = _____

A 3,145

B 3,146

C 3,155

D 9,999

6 Divide: 800 ÷ 20

F 40

G 400

H 4,000

J 40,000

7 Divide: 69 ÷ 3

  A 23   C 36
  B 26   D 66

8 The Hawks scored 68 points in a basketball game. The other team, the Cardinals, scored 49 points. How would you find an estimate for how many more points the Hawks scored than the Cardinals?

  F Subtract 50 from 60.
  G Add 50 and 60.
  H Add 50 and 70.
  J Subtract 50 from 70.

9 Multiply: 6 × 4 × 5

  A 29
  B 120
  C 125
  D 140

10 87 × 9 = _____

  F 135
  G 623
  H 686
  J 783

Short-Response Question

11 Leslie wrote this division sentence in her notebook.

  36 ÷ 9 = 4

Part A

Write a multiplication fact that is related to the division fact.

Answer _____

Part B

Explain how you found your answer.

Lesson 23

Using Open Sentences

Math Words to Know

number sentence a math sentence written with numbers, operation signs ($+, -, \div, \times$), and comparing symbols ($<, >, =$)

open sentence a number sentence that is missing one number

Example 1

Write "four times six equals _____" as an open sentence.

Step 1 Write the numbers as numerals.

four = 4

six = 6

Step 2 Which operation will you use?

times = $\times$

Hint An open sentence has at least two numbers and one operatfion.

Solution $4 \times 6 =$ _____

100 Algebra

Example 2

Dwayne has 64 baseball cards. He gave an equal number of baseball cards to 4 friends. Write an open sentence that he can use to find how many baseball cards each friend will get.

Step 1 Which operation (addition, subtraction, multiplication, or division) will you use to solve the word problem?

You will use division.

Hint Anytime objects are separated or put in equal groups, you are dividing.

Step 2 What numbers will you divide?

64 baseball cards need to be divided among 4 people.

You will divide 64 by 4, or $64 \div 4$.

Hint Numbers must be in the correct order when subtracting and dividing.

Solution $64 \div 4 =$ _____

Example 3

Write a problem that matches this open sentence.

$5 \times 8 =$ _____

Step 1 Which operation will you use to write the problem?

You need to use multiplication.

Step 2 Think of a situation where you need to multiply.

You could think of how many fish are in some fish bowls.

There could be 5 fish bowls with 8 fish in each.

Hint Multiplication can be used when there are a lot of equal groups.

Step 3 Write the problem.

Include all the facts.

Solution There are 5 fish bowls. Each one has 8 fish in it. How many fish are there in all?

Multiple-Choice Questions

1. Brooke divided 72 pieces of chalk into 8 equal groups. Which open sentence can be used to find the number of pieces of chalk in each group?

    A. $72 - 8 =$ _____

    B. $72 \times 8 =$ _____

    C. $72 \div 8 =$ _____

    D. $72 + 8 =$ _____

    Test Taking Tip Which operation should you use?

    You do not need to subtract. Cross it out.

    You do not need to multiply. Cross it out.

    You need to divide. This could be the answer.

    You do not need to add. Cross it out.

2. Which shows "thirty-nine added to fifty-three equals _____?"

    A. $39 \times 53 =$ _____

    B. $39 - 53 =$ _____

    C. $39 \div 53 =$ _____

    D. $39 + 53 =$ _____

3. Which shows "nine multiplied by six equals _____?"

    A. $9 \times 6 =$ _____

    B. $9 + 6 =$ _____

    C. $9 \div 6 =$ _____

    D. $9 - 6 =$ _____

4. Which shows "eighty divided by ten equals _____?"

    A. $80 - 10 =$ _____

    B. $80 \div 10 =$ _____

    C. $10 \div 80 =$ _____

    D. $80 \times 10 =$ _____

5. Which shows "twenty-four subtracted from sixty-seven equals _____?"

    A. $24 + 67 =$ _____

    B. $24 - 67 =$ _____

    C. $67 \div 24 =$ _____

    D. $67 - 24 =$ _____

6. Daryl bought 4 packs of football cards. Each pack has 7 cards. Which open sentence can be used to find the number football cards Daryl bought?

   A. $4 + 7 =$ _____

   B. $7 \div 4 =$ _____

   C. $4 \times 7 =$ _____

   D. $4 \times 4 =$ _____

7. Sandra is reading a magazine with 90 pages. She read 34 pages. Which open sentence can be used to find how many pages Sandra has left to read?

   A. $34 - 90 =$ _____

   B. $90 + 34 =$ _____

   C. $90 - 34 =$ _____

   D. $90 \div 34 =$ _____

8. James spent $18 on tires for his bicycle. He also spent $21 on lights and oil for his bicycle. Which open sentence can be used to find how much James spent on his bike all together?

   A. $18 \times 21 =$ _____

   B. $21 + 18 =$ _____

   C. $18 \div 21 =$ _____

   D. $21 - 18 =$ _____

9. Which problem matches this open sentence?

   $31 + 15 =$ _____

   A. Paula picked 31 apples and 15 pears. How many pieces of fruit did she pick?

   B. Paula picked 31 apples and 15 pears. How many more apples than pears did she pick?

   C. Paula picked 31 apples and 15 pears today. How much fruit did she pick last week?

   D. Paula picked 31 apples and 15 pears. How many oranges did she pick?

Short-Response Question

10. Write a problem that matches this open sentence.

    $8 \times 3 =$ _____

    _____

    _____

    _____

Lesson 24

Comparing Whole Numbers Using Symbols

Math Words to Know

not equal to shown with the symbol ≠

Example 1

Which of these symbols will make the sentence below true: >, =, ≠?

1,395 _____ 1,407

Step 1 Draw a number line to help you compare.

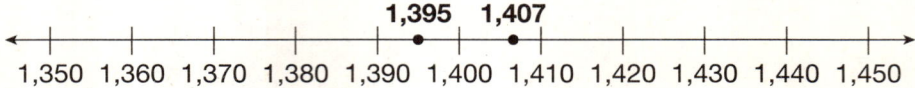

Step 2 1,395 is less than 1,407.

The greater than (>) and equal to (=) symbols are not right.

Hint Cross out answer choices that don't make sense.

The ≠ symbol makes the sentence true.

Solution 1,395 ≠ 1,407

Example 2

Which of these numbers will make the sentence true: 2,699; 3,971; 3,840?

_____ > 3,926

Step 1 What are you asked to find?

You are asked to find which number is **greater than** 3,926.

Step 2 Draw a number line to help you order the numbers.

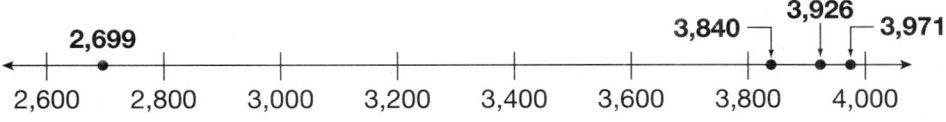

Hint Numbers that are greater are to the right on a number line. Numbers that are lesser are to the left.

Step 3 Which number is to the right of 3,926 on the number line?

3,971 is to the right of 3,926.

Solution 3,971 > 3,926

Multiple-Choice Questions

1. Which symbol will make this sentence true?

    5,065 _____ 5,100

    Test Taking Tip Cross out symbols that do not make sense.

    A. >   This is a comparing symbol. This could be right.
    B. <   This is a comparing symbol. This could be right.
    C. −   This is not a comparing symbol. Cross it out.
    D. +   This is not a comparing symbol. Cross it out.

2. Which of these symbols will make this sentence true?

    8,439 _____ 8,438

    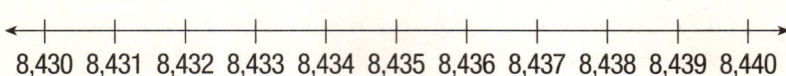

    A. <
    B. =
    C. −
    D. ≠

3. Which number will make this sentence true?

    7,342 < _____

    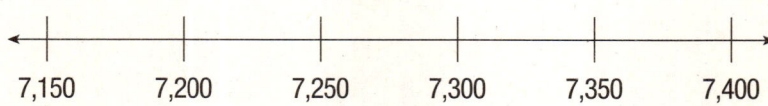

    A. 7,175
    B. 7,246
    C. 7,312
    D. 7,389

106   Algebra

4. Which symbol will make this sentence true?

   2,936 _____ 2,936

   A. <
   B. >
   C. =
   D. ≠

5. Which symbol will make this sentence true?

   3,006 _____ 3,060

   A. <
   B. >
   C. =
   D. +

6. Which number will make this sentence true?

   9,341 < _____

   A. 8,796
   B. 9,302
   C. 9,337
   D. 9,348

7. Which number will make this sentence true?

   _____ < 4,642

   A. 4,624
   B. 4,649
   C. 4,651
   D. 4,700

Short-Response Questions

8. 564 _____ 583

   Part A Draw a number line that can be used to compare the numbers above.

   Part B Write two symbols that will make the number sentence true.

   Answer _____ and _____

9. 6,275 > _____

   Write three numbers that make the number sentence above true.

   _____

Lesson 24: Comparing Whole Numbers Using Symbols

Lesson 25

Geometric and Number Patterns

Math Words to Know

pattern   a series of numbers or shapes that follow a rule

pattern rule   an operation performed on numbers, or an action performed on shapes, in a pattern

Example 1

What is the rule for this pattern?

> 3, 7, 11, 15, 19,…

Step 1   Look at the numbers. Do they increase or decrease?

The numbers increase.

Step 2   Look at the first two numbers. How do they increase?

4 is added to the first number (3) to get the second number (7).

Step 3   Look at the second and third numbers. Do they increase by the same amount?

Yes, the next two numbers increase by 4.

Hint The rule for a pattern must apply to every number in the pattern.

Step 4   Continue to look at the rest of the numbers. Do they all increase by 4?

Yes, 4 is added to each number.

Solution   The rule for the pattern is to add 4.

Example 2

What number comes next in this pattern?

16, 13, 10, 7, _____, ...

Step 1 Look at the numbers. Do they increase or decrease?

The numbers decrease.

Step 2 How do the numbers decrease?

They decrease by 3.

Step 3 Does the rule work for all the numbers in the pattern?

Yes, they all decrease by 3.

Step 4 Apply the rule to the last number in the pattern.

7 − 3 = 4

Solution The next number in the pattern is 4.

Example 3

What is the missing number in this pattern?

37, 42, 47, 52, _____, 62, 67

Step 1 Look at the numbers. Do they increase or decrease?

The numbers increase.

Step 2 What is the rule for the pattern?

Add 5.

Step 3 Apply the rule to the number before the blank in the pattern.

52 + 5 = 57

Hint Apply the rule to the number after the blank to check your answer. You should get 62.

Solution The missing number is 57.

Multiple-Choice Questions

1. What is the rule for this pattern?

    98, 91, 84, 77, 70,…

		Test Taking Tip Are the numbers in the pattern increasing or decreasing?
A.	add 7	The numbers are not increasing. Cross it out.
B.	multiply by 7	The numbers are not increasing. Cross it out.
C.	subtract 9	The numbers are decreasing. This could be right.
D.	subtract 8	The numbers are decreasing. This could be right.

2. What is the rule for this pattern?

    9, 21, 33, 45, 57,…

    A. multiply by 3
    B. add 13
    C. subtract 12
    D. add 12

3. What is the rule for this pattern?

    77, 67, 57, 47, 37,…

    A. subtract 10
    B. subtract 5
    C. subtract 9
    D. add 10

4. What number comes next in this pattern?

    0, 6, 12, 18, ____

    A. 21
    B. 22
    C. 23
    D. 24

5. What number comes next in this pattern?

    35, 44, 53, 62, 71, ____

    A. 78
    B. 79
    C. 80
    D. 82

6. What number comes next in this pattern?

   68, 57, 46, 35, _____

   A. 24
   B. 25
   C. 26
   D. 34

7. What is the missing number in this pattern?

   1, 9, 17, _____, 33, 41

   A. 24   C. 26
   B. 25   D. 28

8. What is the missing number in this pattern?

   41, 39, 37, 35, _____, 31

   A. 30
   B. 32
   C. 33
   D. 34

9. What is the missing number in this pattern?

   21, _____, 13, 9, 5, 1

   A. 15   C. 17
   B. 16   D. 18

Short-Response Questions

10. Donald wrote this pattern on the board.

   54, 61, 68, 75, _____

   Part A What is the rule for this pattern?
   Answer _____

   Part B What is the next number in this pattern?
   Answer _____

11. What is the missing number in this pattern?

   38, 33, 28, 23, _____, 13

   Answer _____

Lesson 25: Geometric and Number Patterns   111

Lesson 26

Finding the Rule for an Input-Output Table

Math Words to Know

input-output table   a table that uses a rule to change an input number to an output number

Example 1

For this input-output table, the rule is *subtract 5*. Complete the table.

Input Numbers	Output Numbers
10	?
11	?
12	7

Step 1   Find the first input number.

The first input number is 10.

Apply the rule *subtract 5* to 10.

$10 - 5 = \mathbf{5}$

Step 2   Find the second input number.

The second input number is 11.

Apply the rule *subtract 5* to 11.

$11 - 5 = \mathbf{6}$

Hint   The rule for an input-output table is the same for each input number.

Solution   The completed input-output table with the rule *subtract 5*:

Input Numbers	Output Numbers
10	5
11	6
12	7

Example 2

For this input-output table, the rule is *add 8*. Find the missing numbers.

Input Numbers	Output Numbers
?	11
?	15
11	19
15	?

Step 1 Find the first input number.

What number, added to 8, equals 11?

3 + 8 = 11

Step 2 Find the second input number.

What number, added to 8, equals 15?

7 + 8 = 15

Step 3 Use the rule to find the missing output number.

15 + 8 = **23**

Solution Here is the completed input-output table with the rule *add 8*.

Input Numbers	Output Numbers
3	11
7	15
11	19
15	23

The missing input numbers are 3 and 7. The missing output number is 23.

Multiple-Choice Questions

1. Find the missing output numbers if the rule is *add 1*.

Input Numbers	Output Numbers
1	?
2	?
3	?

   Test Taking Tip Should output numbers be less than, equal to, or greater than the input numbers if you **add 1**?

   A. 0, 1, 2   These numbers are less than the input numbers. Cross it out.

   B. 1, 2, 3   These numbers are equal to the input numbers. Cross it out.

   C. 2, 3, 4   These numbers are greater than the input numbers. This could be right.

   D. 4, 5, 6   These numbers are greater than the input numbers. This could be right.

2. Find the missing output numbers if the rule is *subtract 4*.

Input Numbers	Output Numbers
18	?
25	?
30	?

   A. 22, 29, 34

   B. 14, 22, 26

   C. 15, 21, 27

   D. 14, 21, 26

3. Find the missing output numbers if the rule is *add 16*.

Input Numbers	Output Numbers
20	?
35	?
41	?

   A. 36, 41, 56

   B. 4, 19, 25

   C. 36, 51, 57

   D. 36, 31, 57

114   Algebra

4. Find the missing input numbers if the rule is *add 3*.

Input Numbers	Output Numbers
7	10
?	11
?	12
?	13

A. 8, 9, 10
B. 14, 15, 16
C. 9, 10, 11
D. 13, 14, 15

5. Find the missing input numbers if the rule is *subtract 7*.

Input Numbers	Output Numbers
?	7
?	10
?	14

A. 0, 3, 7
B. 14, 17, 21
C. 0, 7, 14
D. 10, 14, 21

Short-Response Questions

6. Find the missing output numbers if the rule is *subtract 2*.

Input Numbers	Output Numbers
34	
35	
36	

7. Find the missing input numbers if the rule is *add 9*.

Input Numbers	Output Numbers
	16
	23
	28

Lesson 26: Finding the Rule for an Input-Output Table 115

PROGRESS CHECK 1

1 Which shows "forty-six added to nineteen equals ___?"

    A   46 + 9 = ___

    B   46 × 19 = ___

    C   46 + 19 = ___

    D   46 − 19 = ___

2 Which shows "fifty-four divided by six equals ___?"

    F   54 − 6 = ___

    G   54 + 6 = ___

    H   54 × 6 = ___

    J   54 ÷ 6 = ___

3 Which of the symbols will make this sentence true?

    2,531 ___ 2,533

    A   >

    B   =

    C   ≠

    D   −

4 Which symbol will make this sentence true?

    9,476 ___ 9,426

    F   <

    G   >

    H   =

    J   +

5 What is the rule for this pattern?

    57, 55, 53, 51,…

    A   subtract 2

    B   add 2

    C   divide by 2

    D   subtract 3

6 What is the rule for this pattern?

    8, 15, 22, 29, 36,…

    F   add 9

    G   subtract 7

    H   add 8

    J   add 7

7 What is the rule for this input-output table?

Input Numbers	Output Numbers
11	19
13	21
15	23
17	25

- **A** subtract 8
- **B** add 2
- **C** subtract 2
- **D** add 8

8 What is the rule for this input-output table?

Input Numbers	Output Numbers
10	7
9	6
8	5
7	4

- **F** subtract 3
- **G** subtract 1
- **H** add 3
- **J** add 1

Short-Response Question

9 Ms. Lind wrote the pattern below on the board.

42, 33, 24, 15, ____

Part A

What is the rule for the pattern?

Answer _____

Part B

What is the missing number in the pattern?

Answer _____

PROGRESS CHECK 2

1 Kelly saved $23. She spent $9 on a hat and gloves. Which open sentence can be used to find how much money Kelly has left?

  A 23 − 9 = ____

  B 23 + 9 = ____

  C 23 × 9 = ____

  D 23 ÷ 9 = ____

2 Mr. Shu works 8 hours each day. Which open sentence can be used to find the number of hours Mr. Shu works in 6 days?

  F 8 + 6 = ____

  G 8 × 6 = ____

  H 8 − 6 = ____

  J 8 ÷ 6 = ____

3 Gina has 47 arcade tickets. She won 81 more. Which open sentence can be used to find the total number of tickets Gina has?

  A 81 ÷ 47 = ____

  B 47 × 81 = ____

  C 81 − 47 = ____

  D 47 + 81 = ____

4 Which number will make this sentence true?

  4,814 < ____

  F 3,956   H 4,805

  G 4,729   J 4,823

5 Which number will make this sentence true?

  ____ > 6,389

  A 5,980

  B 6,217

  C 6,345

  D 6,390

6 Which number comes next in this pattern?

  75, 79, 83, 87, ____

  F 89

  G 90

  H 91

  J 92

7 What is the missing number in this pattern?

  34, 29, 24, 19, ____, 9, 4

  A 13

  B 14

  C 15

  D 16

8 Find the missing input numbers if the rule is *add 6*.

Input Numbers	Output Numbers
2	8
	10
	12
	14

F 4, 6, 8   H 4, 5, 6
G 16, 18, 20   J 3, 4, 5

9 Find the missing output numbers if the rule is *subtract 10*.

Input Numbers	Output Numbers
21	
23	
25	

A 31, 33, 35   C 11, 13, 15
B 11, 21, 31   D 31, 21, 11

Short-Response Question

10 Look at this pattern.

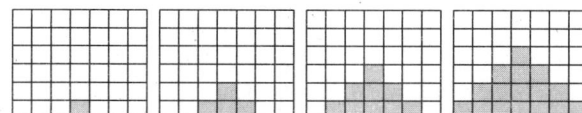

Part A

Shade the squares in this figure to show the next figure in the pattern.

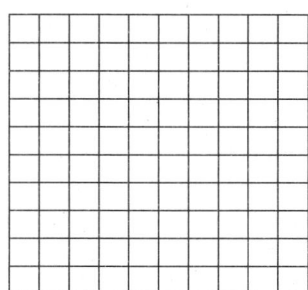

Part B

Tell how you found your answer to Part A.

_____

_____

_____

Lesson 27

Two-Dimensional Figures

Math Words to Know

line segment   a part of a line that has two endpoints

ray   part of a line that has one endpoint and goes on forever in one direction

angle   a figure that is made up of two rays or line segments with the same endpoint

closed figure   a shape with no openings or gaps

polygon   a closed figure made up of three or more line segments

triangle   a polygon with 3 sides and 3 angles

quadrilateral   a polygon with 4 sides and 4 angles; examples include a square, rectangle, parallelogram, and rhombus

pentagon   a polygon with 5 sides and 5 angles

hexagon   a polygon with 6 sides and 6 angles

octagon   a polygon with 8 sides and 8 angles

Example 1

How many sides and angles are there in this figure?

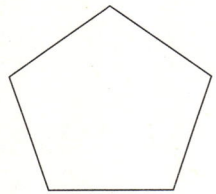

Step 1   Count the number of sides.

Five line segments make up 5 sides of the figure.

Step 2   Count the number of angles.

There are 5 angles in the figure.

Hint   Polygons have the same number of angles and sides.

Solution   The figure has five sides and five angles.

Example 2

What is the name of this geometric figure?

•⎯⎯⎯⎯→

Step 1 Count the number of endpoints.

There is one endpoint.

Step 2 If the figure has one endpoint it is either a ray or an angle.

Step 3 Count the number of line segments or rays.

There is one ray. It has an endpoint and extends forever in one direction.

Solution The figure is a ray.

Example 3

Is this figure a polygon? If so, what is the name of this figure?

Step 1 A polygon is a closed figure. Is this a closed figure?

Yes, the figure is closed. It is a polygon.

Step 2 Count the number of sides and angles

The polygon has 4 sides and 4 angles.

Hint Use the definitions on the last page to help you identify the polygon. Remember that polygons do not have any curved sides.

Step 3 What type of polygon has 4 sides and 4 angles?

A quadrilateral has 4 sides and 4 angles.

Solution The figure is a polygon. It is a quadrilateral.

Lesson 27: Two-Dimensional Figures   121

Multiple-Choice Questions

1. What is the name of this figure?

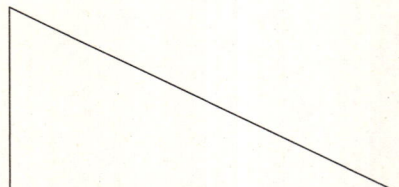

		Test Taking Tip Cross out answers that are not polygons.
A.	line segment	This is not the name of a polygon. Cross it out.
B.	triangle	This is the name of a polygon. This could be right.
C.	angle	This is not the name of a polygon. Cross it out.
D.	quadrilateral	This is the name of a polygon. This could be right.

2. How many sides does this figure have?

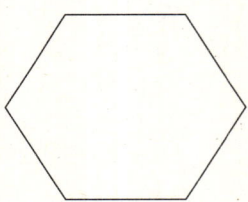

- A. 4
- B. 5
- C. 6
- D. 7

3. How many angles does this figure have?

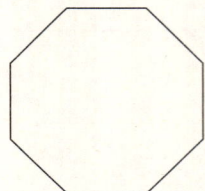

- A. 6
- B. 7
- C. 8
- D. 9

4. Which of the following figures is **not** a polygon?

A.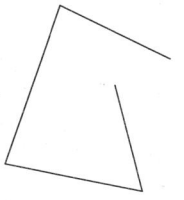

B.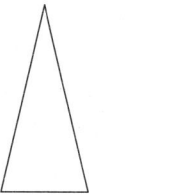

C.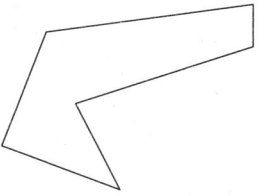

D.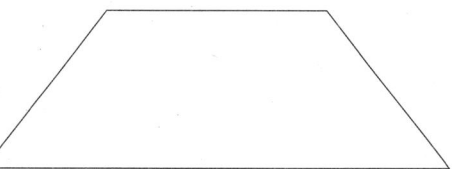

5. What is the name of this figure?

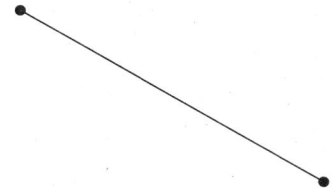

A. ray

B. angle

C. polygon

D. line segment

Short-Response Question

6. Barry drew the figure below.

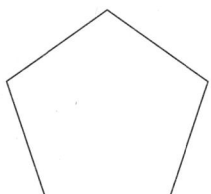

Part A How many angles does this figure have?

Answer _____ angles

Part B What is the name of this figure?

Answer _____

Lesson 28

Perimeter and Area

Math Words to Know

perimeter   the measure of the distance around a figure

area   the number of square units needed to cover a figure

Example 1

What is the perimeter of the hexagon?

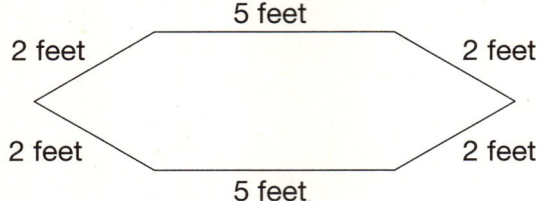

Step 1   Write the sum of the lengths of each side as an open addition sentence.

5 + 5 + 2 + 2 + 2 + 2 = _____

Hint Make sure you include the lengths for **all** the sides, even if some sides have the same length.

Step 2   Add.

5 + 5 + 2 + 2 + 2 + 2 = 18

Step 3   Label the answer with units.

18 feet

Solution   The perimeter of the hexagon is 18 feet.

124   Geometry

Example 2

What is the area of this rectangle?

Step 1 Count the number of square units inside the rectangle.

There are 20 units.

Step 2 Label the units.

20 square units

Hint Make sure you label the units when finding the area and perimeter of figures.

Solution The area of the rectangle is 20 square units.

Multiple-Choice Questions

1. What is the perimeter of this rectangle?

```
         18 units
3 units [          ] 3 units
         18 units
```

Test Taking Tip Estimate to check your answer.

A.	21 units	21 units is close to the length of just one side. Cross it out.
B.	42 units	This measure is close to the estimate. This could be right.
C.	45 units	This measure is close to the estimate. This could be right.
D.	54 units	This measure is too large. Cross it out.

2. What is the perimeter of this triangle?

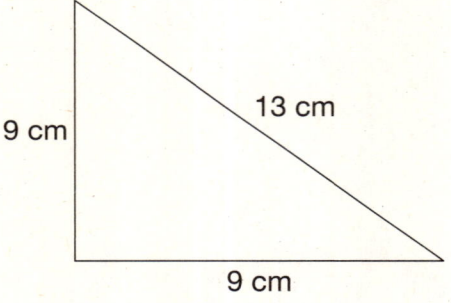

- **A.** 18 cm
- **B.** 22 cm
- **C.** 31 cm
- **D.** 40 cm

3. What is the perimeter of this figure?

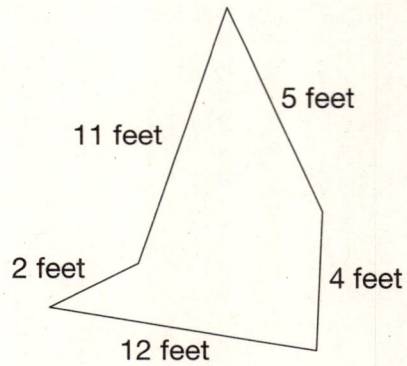

- **A.** 23 feet
- **B.** 29 feet
- **C.** 32 feet
- **D.** 34 feet

126 Geometry

4. What is the area of this rectangle?

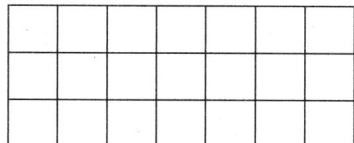

- A. 20 square units
- B. 21 square units
- C. 22 square units
- D. 24 square units

5. What is the area of this rectangle?

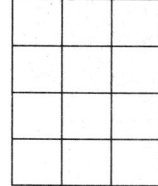

- A. 12 square units
- B. 14 square units
- C. 16 square units
- D. 28 square units

Short-Response Questions

6. Kate drew the figure below.

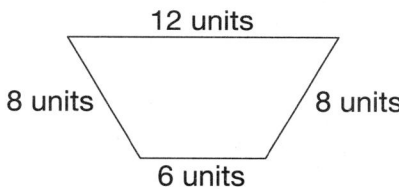

What is the perimeter of this figure?

Answer _____

7. Doug drew the square below.

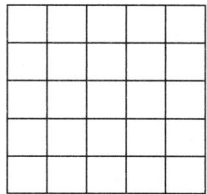

What is the area of the square?

Answer _____

Lesson 29

Three-Dimensional Figures

Math Words to Know

face a flat side of a three-dimensional figure

edge a line segment where two faces meet on a three-dimensional figure

vertex a point where three or more edges meet on a three-dimensional figure

prism a three-dimensional figure with matching bases that are polygons

cube a three-dimensional figure with faces that are squares

pyramid a three-dimensional figure with a base that is a polygon; the other faces are triangles

cylinder a three-dimensional figure with matching bases that are circles; the figure has no edges or vertices

cone a pointed, three-dimensional figure with a circular base

sphere a three-dimensional figure with no faces, edges, or vertices

Example 1

What is the name of this figure?

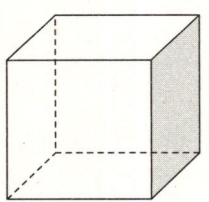

Step 1 Describe the figure.

The figure is three-dimensional.

The faces of the figure are the same size and shape.

The faces are squares.

Step 2 Name figures that have those features.

prism, cube

Hint Sometimes a figure has more than one name.

Solution The figure is a prism and a cube.

Example 2

How many faces does this figure have?

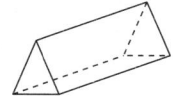

Step 1 Count the number of faces shaped like triangles.

There are 2 faces shaped like triangles.

Step 2 Count the number of faces shaped like rectangles.

There are 3 faces shaped like rectangles.

Step 3 Add the number of faces that you counted.

2 + 3 = 5

Solution The figure has 5 faces.

Example 3

What shape are the faces of this figure?

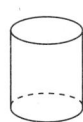

Step 1 Count the number of faces in the figure.

There are 2 faces.

Step 2 What type of shape are the faces.

Both faces are circles.

Hint A face is flat, so the curved part of the figure is not a face.

Solution The figure has 2 circles for faces.

Multiple-Choice Questions

1. What is the name of this figure?

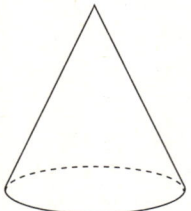

   Test Taking Tip Cross out answers that are not three-dimensional figures.

   A. circle — This is not a three-dimensional figure. Cross it out.
   B. cylinder — This is a three-dimensional figure. This could be right.
   C. oval — This is not a three-dimensional figure. Cross it out.
   D. cone — This is a three-dimensional figure. This could be right.

2. What is the name of this figure?

   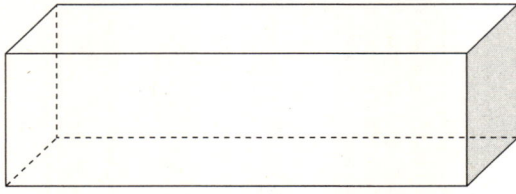

   A. rectangular prism
   B. triangular prism
   C. cube
   D. rectangular pyramid

3. How many faces does this figure have?

   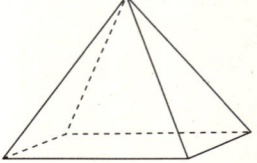

   A. 4
   B. 5
   C. 6
   D. 7

4. How many faces does this figure have?

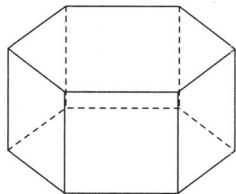

- A. 6
- B. 7
- C. 8
- D. 9

5. What are the faces of this triangular pyramid?

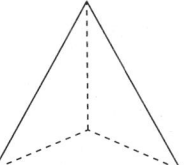

- A. 1 rectangle, 3 triangles
- B. 1 rectangle, 4 triangles
- C. 3 triangles
- D. 4 triangles

Short-Response Questions

6. What is the name of this figure?

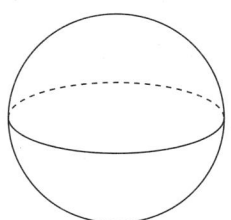

Answer _____

7. Tiffany drew this figure.

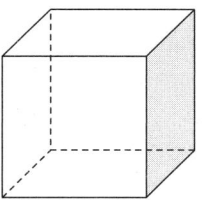

Part A   How many faces does the figure have?

Answer _____ faces

Part B   What is the name of the figure?

Answer _____

Lesson 29: Three-Dimensional Figures

PROGRESS CHECK 1

1 How many sides does this figure have?

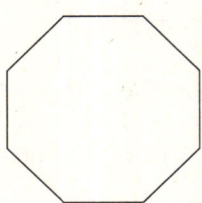

- A 6
- B 7
- C 8
- D 9

2 What is the perimeter of this figure?

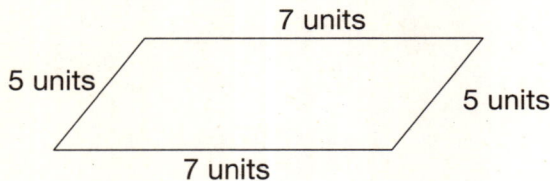

- F 12 units
- G 19 units
- H 24 units
- J 35 units

3 What is the name of this figure?

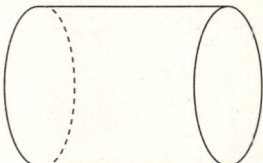

- A sphere
- B cylinder
- C cone
- D circle

4 How many angles does this figure have?

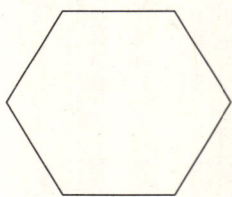

- F 5
- G 6
- H 7
- J 8

5 What is the area of this rectangle?

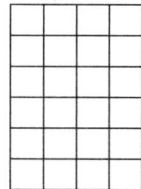

A 10 square units

B 20 square units

C 24 square units

D 28 square units

6 What is the name of this figure?

F ray

G angle

H polygon

J line segment

Short-Response Question

7 Jason drew this figure.

Part A

What is the name of this figure?

Answer _____

Part B

How many sides does this figure have?

Answer _____ sides

Part C

What is the perimeter of the figure?

Answer _____

Progress Check 1   133

PROGRESS CHECK 2

1 What are the faces of this pyramid?

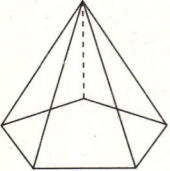

- A 6 triangles
- B 1 square, 5 triangles
- C 1 pentagon, 4 triangles
- D 1 pentagon, 5 triangles

2 Which of the following figures is **not** a polygon?

F
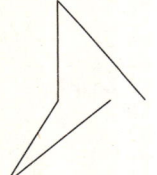

G
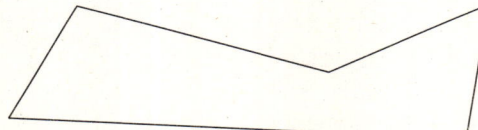

H
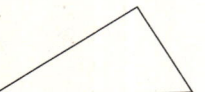

J
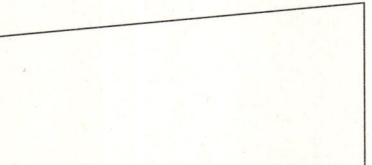

3 What is the area of this rectangle?

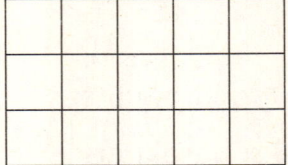

- A 15 square units
- B 16 square units
- C 18 square units
- D 20 square units

4 How many faces does this figure have?

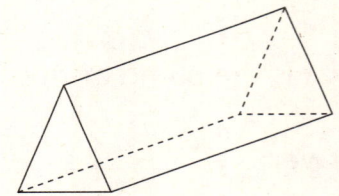

- F 4 faces
- G 5 faces
- H 6 faces
- J 7 faces

5 What is the perimeter of this figure?

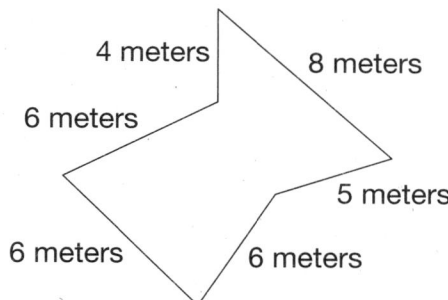

A 28 meters   C 31 meters
B 29 meters   D 35 meters

6 How many vertices does this figure have?

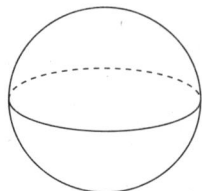

F 0
G 1
H 2
J 4

Short-Response Question

7 Alyssa's kitchen table is a rectangle represented in the picture below.

Part A

What is the area of the rectangle?

Answer _____ square units

Part B

Explain why your answer in Part A is right.

_____

_____

Progress Check 2 135

Lesson 30

Selecting Tools and Units for Measuring Length

Math Words to Know

measure   to find the size or capacity of an object

length   the distance from one point to another

Example 1

Which is the best unit for measuring the distance from New York City to Albany?

>   foot   mile   yard

- **Step 1**   Describe each unit.

    A foot is 12 inches.

    A mile is 5,280 feet.

    A yard is 3 feet.

- **Step 2**   Describe the distance being measured.

    The distance from New York City to Albany is a long distance.

- **Step 3**   Which unit is used for measuring long distances?

    mile

    Hint Even though two or more units can be used for measuring the same distance, one unit is often better than another.

- **Solution**   A mile is the best unit for measuring the distance from New York City to Albany.

Example 2

Which is the best tool for measuring the length of a cat's tail?

    ruler   odometer   yard stick

Step 1 Describe each tool.
    A ruler is a foot long with measurements in inches.
    An odometer is used to measure distances traveled by a car in miles and tenths of miles.
    A yard stick is 3 feet long.

Step 2 What units are usually used to measure the length of a cat's tail?
    inches

Step 3 Which is the best tool for measuring the length of a cat's tail?
    ruler

Solution A ruler is the best tool for measuring the length of a cat's tail.

Example 3

What is the best estimate for the measure of the length of a side of a notebook?

    10 inches   10 feet   10 yards

Step 1 What is the best unit for measuring the length of a notebook?
    centimeters or inches

Step 2 What is a range of lengths for the side of a notebook?
    about 9 to 12 inches

Solution The best estimate for the measure of the length of a side of a notebook is 10 inches.

Multiple-Choice Questions

1. Which is the best unit for measuring the length of a basketball court?

   Test Taking Tip Cross out answers that do not make sense.

   A. inch — This unit is small but possible. This could be right.

   B. centimeter — This unit is too small. Cross it out.

   C. mile — This unit is too large. Cross it out.

   D. yard — This unit is reasonable. This could be right.

2. Which is the best tool for measuring the length of a paper clip?

   A. odometer

   B. yard stick

   C. ruler

   D. scale

3. What is the best estimate for measuring the height of a horse's shoulders?

   A. 5 inches

   B. 5 meters

   C. 5 miles

   D. 5 feet

4. Which is the best unit for measuring the distance from Syracuse to Buffalo?

   A. mile

   B. foot

   C. yard

   D. inch

5. Which is the best tool for measuring the length of a pencil?

   A. meter stick

   B. ruler

   C. odometer

   D. yard stick

6. What is the best estimate for measuring the height of a refrigerator?

   A. 5 miles

   B. 5 inches

   C. 5 feet

   D. 5 yards

7. Which is the best unit for measuring the length of a car?

   A. foot

   B. millimeter

   C. centimeter

   D. mile

8. Which is the best tool for measuring the length of your foot?

   A. odometer

   B. ruler

   C. yard stick

   D. meter stick

9. What is the best estimate for measuring the length of a playground?

   A. 50 centimeters

   B. 50 inches

   C. 50 miles

   D. 50 yards

Short-Response Questions

10. What is the best unit for measuring the height of a basketball hoop?

    Answer _____

11. **Part A**   What is the best tool for measuring the length of a pen?

    Answer _____

    Part B   Estimate the length of a pen.

    Answer _____

Lesson 31

Measuring Length

Math Words to Know

metric system a measurement system using millimeters, centimeters, meters, and kilometers to measure length

customary system a measurement system using inches, feet, and yards to measure length

Example 1

The arrow is pointing to a measurement on this ruler. What is the measurement?

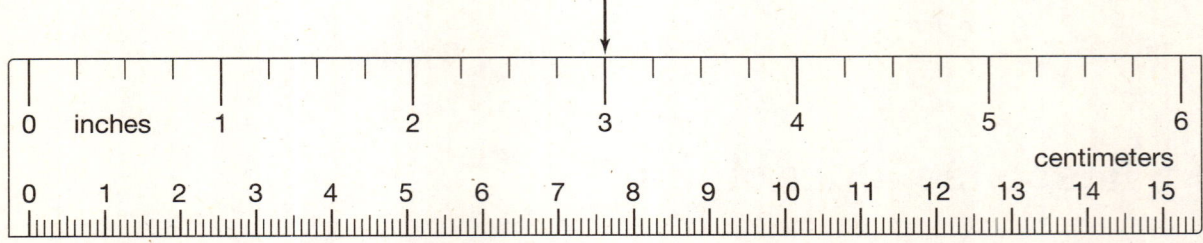

Step 1 What number is the arrow pointing to?

The arrow is pointing to 3.

Step 2 What kind of units is the ruler measuring?

inches

Hint An inch is longer than a centimeter.

Solution The arrow is pointing to 3 inches.

Example 2

What is the length of this eraser to the nearest centimeter?

Step 1 Make sure the left side of the eraser is lined up with the 0 mark on the ruler.

Step 2 Look at the right side of the eraser. What number is below it?

6

Step 3 What kind of units is the ruler measuring?

centimeters

Hint Make sure you use the right units for your measurements.

Solution The length of the eraser is 6 centimeters.

Multiple-Choice Questions

1. The arrow is pointing to a measurement on this ruler. What is the measurement?

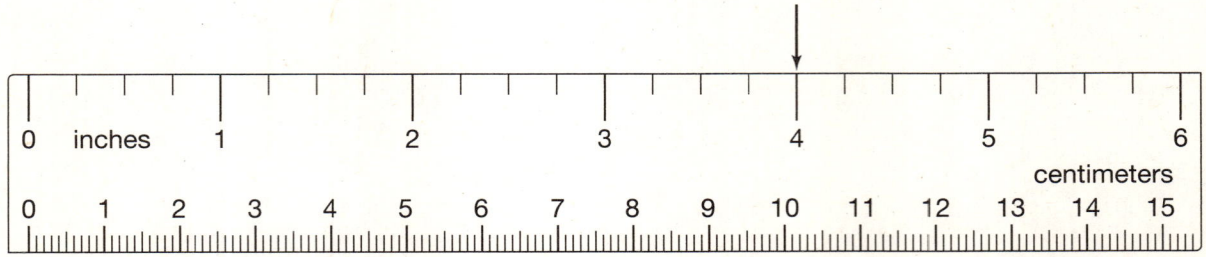

Test Taking Tip Cross out answers are the wrong units.

- **A.** 3 inches — This measurement is in inches. This could be right.
- **B.** 4 inches — This measurement is in inches. This could be right.
- **C.** 4 centimeters — This measurement is in centimeters. Cross it out.
- **D.** 8 centimeters — This measurement is in centimeters. Cross it out.

2. The arrow is pointing to a measurement on this ruler. What is the measurement?

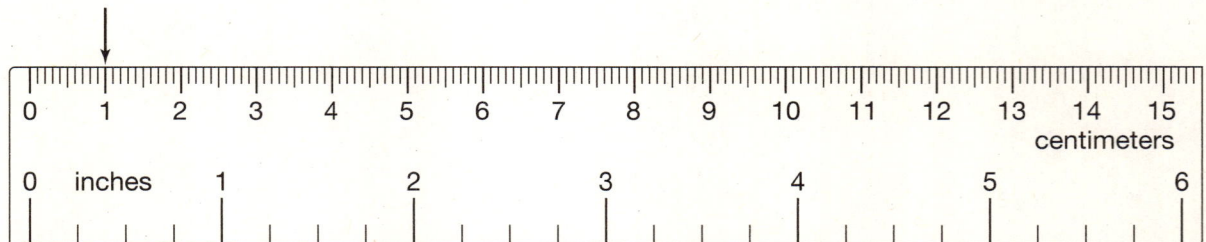

- **A.** 0 inches
- **B.** 1 inch
- **C.** 1 centimeter
- **D.** 2 meters

3. What is the length of this worm to the nearest centimeter?

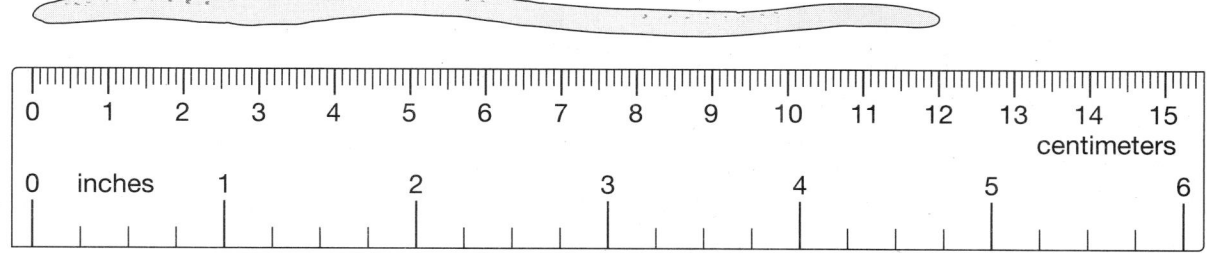

- A. 11 centimeters
- B. 11 inches
- C. 12 inches
- D. 12 centimeters

4. What is the length of this pencil to the nearest inch?

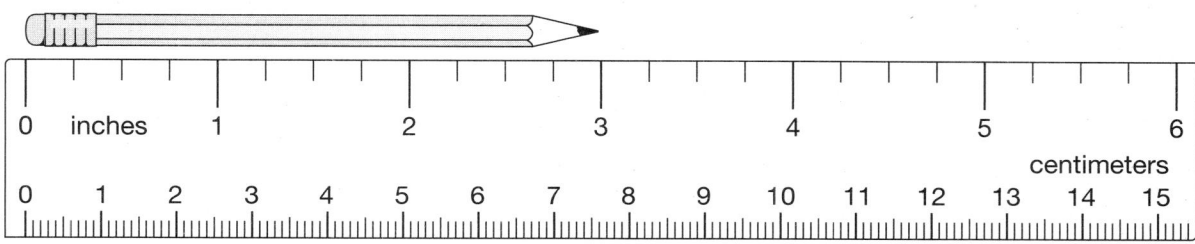

- A. 1 inch
- B. 2 inches
- C. 3 inches
- D. 4 inches

Short-Response Question

5. What is the length of this star to the nearest inch?

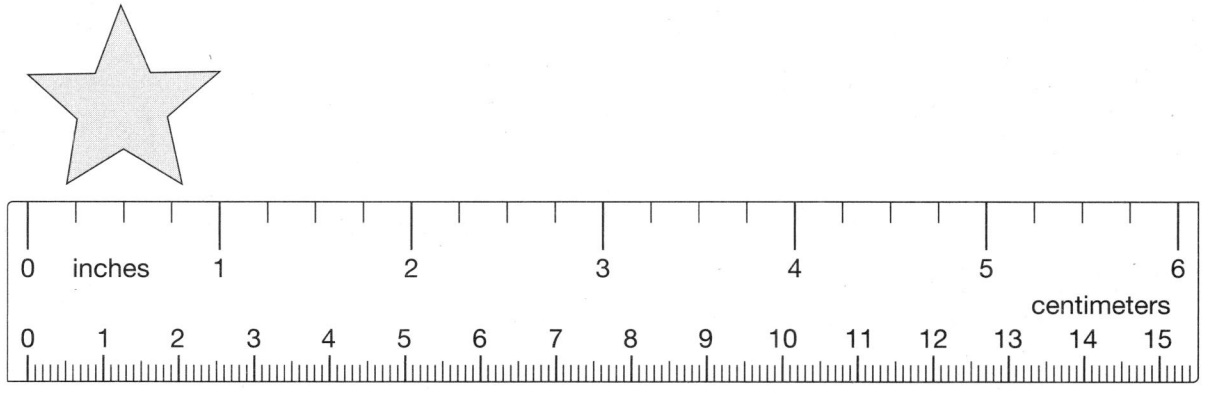

Answer _____

Lesson 31: Measuring Length 143

Lesson 32

Selecting Tools and Units for Measuring Mass

Math Words to Know

mass the amount of matter in an object

Example 1

Which is the best unit for measuring the mass of a pen?

　　　　gram　　　　kilogram　　　　inch

Step 1 Describe each unit.

A gram is used for measuring the mass of small objects. A paperclip weighs about 1 gram.

A kilogram is used for measuring the mass of larger objects. The mass of 1 liter of water is 1 kilogram.

An inch is used to measure length, not mass.

Step 2 Describe the object that is being measured.

A pen is a small object.

Hint When choosing the best unit to measure mass, be sure not to choose units that are too large or too small.

Step 3 Which unit is used for measuring small objects?

gram

Solution A gram is the best unit for measuring the mass of a pen.

Example 2

A scale shows a mass of 5 kilograms. Which object is most likely on the scale?

    laptop computer   house   CD

Step 1 Describe the mass of each object.

          A laptop computer weighs a few kilograms.

          A house weighs thousands of kilograms.

          A CD weighs about 5 grams.

Step 2 Which object weighs closest to 5 kilograms?

          the laptop computer

Solution The laptop computer is most likely on the scale.

Example 3

Which is the best estimate for the mass of an apple?

    150 grams   150 kilograms   150 feet

Step 1 What is the best unit for measuring the mass of an apple?

          gram

Step 2 What is a range of masses for an apple?

          about 100 to 200 grams

Solution The best estimate for the mass of an apple is about 150 grams.

Multiple-Choice Questions

1. Which is the best estimate for the mass of a watch?

   Test Taking Tip Cross out answers that do not make sense.

   A. 5 grams   This mass is small but reasonable. This could be right.
   B. 50 grams   This mass is reasonable. This could be right.
   C. 5 kilograms   This mass is too large. Cross it out.
   D. 50 kilograms   This mass is too large. Cross it out.

2. Which is the best unit for measuring the mass of a tennis ball?

   A. gram
   B. meter
   C. kilogram
   D. inch

3. A scale shows a mass of 30 kilograms. Which object is most likely on the scale?

   A. notebook
   B. car
   C. puppy
   D. bicycle

4. Which is the best estimate for the mass of a cell phone?

   A. 40 grams
   B. 400 grams
   C. 4 kilograms
   D. 40 kilograms

5. Which is the best unit for measuring the mass of a toothbrush?

   A. kilogram
   B. inch
   C. gram
   D. foot

6. A scale shows a mass of 1 gram. Which object is most likely on the scale?

   A. dime
   B. stapler
   C. trumpet
   D. shoe

7. Which is the best estimate for the mass of a desk chair?

   A. 5 grams
   B. 50 grams
   C. 500 grams
   D. 5 kilograms

8. Which is the best unit for measuring the mass of a truck?

   A. mile
   B. gram
   C. kilogram
   D. yard

9. A scale shows a mass of 100 grams. Which object is most likely on the scale?

   A. fish hook
   B. scissors
   C. peanut
   D. brick

Short-Response Question

10. Alysha is measuring the mass of a horse.

    What is the best unit for measuring the mass of a horse?

    Answer _____

Lesson 33

Selecting Tools and Units for Measuring Capacity

Math Words to Know

capacity   the measure of how much a container can hold

Example 1

Which is the best unit for measuring the capacity of a bathtub?

    milliliter   liter   kilogram

Step 1 Describe each unit.

A milliliter is used for measuring the capacity of small containers. A drop of water is about 1 milliliter.

A liter is used for measuring the capacity of larger containers. A sports bottle holds about 1 liter.

A kilogram is a unit of mass, not capacity.

Step 2 Describe the container that is being measured.

A bathtub is a large container.

Hint When choosing the best unit to measure capacity, be sure not to choose units that are too small or too large.

Step 3 Which unit is best used for measuring the capacity of larger containers?

liter

Solution   A liter is the best unit for measuring the capacity of a bathtub.

148   Measurement

Example 2

A container holds 200 milliliters. Which container is it most likely to be?

      sink   thimble   drinking glass

Step 1 Determine how large or small 200 milliliters is.

      200 milliliters = 0.2 liters. It is a small capacity.

Step 2 Describe the capacity of each object given.

      The capacity of a sink is large.

      The capacity of a thimble is very small.

      The capacity of a drinking glass is small.

Solution The drinking glass is most likely the container that holds 200 milliliters.

Example 3

Which is the best estimate for the capacity of an eye dropper?

      15 milliliters   15 liters   15 kilometers

Step 1 What is the best unit for measuring the capacity of an eye dropper?

      An eye dropper is very small.

      Milliliters are the best unit for measuring the capacity of very small objects.

Step 2 What is a range of capacities for an eye dropper?

      about 10 to 20 milliliters

Solution The best estimate for the capacity of an eye dropper is 15 milliliters.

Multiple-Choice Questions

1. What is the best estimate for the capacity of a water balloon?

   Test Taking Tip Picture in your mind the amount of liquid a balloon can hold.

   A. 1 milliliter   This capacity is too small. Cross it out.

   B. 10 milliliters   This capacity is too small. Cross it out.

   C. 1 liter   This capacity is reasonable. This could be the right answer.

   D. 10 liters   This capacity is too large. Cross it out.

2. Which is the best unit for measuring the capacity of a juice box?

   A. milliliter

   B. gram

   C. liter

   D. kilogram

3. A container holds 6 liters. What is the container most likely to be?

   A. drinking glass

   B. water bottle

   C. bucket

   D. eye dropper

4. What is the best estimate for the capacity of a bathtub?

   A. 100 milliliters

   B. 10 liters

   C. 400 milliliters

   D. 400 liters

5. Which is the best unit for measuring the capacity of a coffee mug?

   A. liter

   B. kilogram

   C. gram

   D. milliliter

6. A container holds 2 liters. What is the container most likely to be?

   A. eye dropper

   B. thimble

   C. juice box

   D. fish bowl

7. What is the best estimate for the capacity of ink in a pen?

   A. 50 milliliters

   B. 5 liters

   C. 50 liters

   D. 500 milliliters

8. Which is the best unit for measuring the capacity of a soup can?

   A. liter

   B. milliliter

   C. kilogram

   D. gram

9. An object holds 15 milliliters of a liquid. What is the object most likely to be?

   A. frying pan

   B. drinking glass

   C. spoon

   D. sports bottle

Short-Response Question

10. Jerry is filling an ice cube tray with water.

    What is the best unit for measuring the capacity of an ice cube tray?

    Answer _____

Lesson 34

Making Change

Math Words to Know

dollar   a unit of money equal to 100 cents

Example 1

What is the value of the coins shown?

quarter   quarter   dime   nickel   penny   penny

Step 1 Start with the quarters. What is the value of the quarters?

A quarter equals 25 cents.

two quarters = 25 + 25 = 50 cents

Step 2 Add the dime to your total.

A dime equals 10 cents.

50 + 10 = 60 cents

Step 3 Add the nickel to your total.

A nickel equals 5 cents.

60 + 5 = 65 cents

Step 4 Add the pennies to your total.

A penny equals 1 cent.

65 + 2 = 67 cents

Hint Continue to add the coins until you have added them all.

Solution The value of the coins is 67 cents.

152   Measurement

Example 2

Rick bought a pack of gum for 55 cents. He paid with 75 cents. What coins could he get back as change?

Step 1 Decide what you will do with the amounts.

When finding change, subtract the amounts.

Step 2 Subtract.

75 − 55 = 20

Step 3 Use coins to show the amount.

You can show 20 cents with 2 dimes.

Hint You can use different coins to show the same amount. For example, you can also use 4 nickels to show 20 cents.

Solution Rick could get back 2 dimes as change.

Example 3

Latisha bought a clip for her hair for 89 cents. She paid with 1 dollar. How much change did she get back?

Step 1 Decide what you will do with the amounts.

When finding change, you subtract the amounts.

Step 2 What is the value of a dollar in cents?

One dollar equals 100 cents.

Step 3 Subtract.

100 − 89 = 11

Solution Latisha got back 11 cents in change.

Multiple-Choice Questions

1. What is the value of the coins shown?

quarter quarter quarter nickel penny penny penny penny penny penny

Test Taking Tip Add the quarters first. Then add the nickel and pennies.

A.	81 cents	This amount is not enough. Cross it out.
B.	85 cents	This amount is close. This could be right.
C.	86 cents	This amount is close. This could be right.
D.	91 cents	This amount is too much. Cross it out.

2. David bought a rubber ball for 38 cents. He paid with 50 cents. Which coins should he get back as the right change?

   A. dime dime penny penny

   B. nickel penny penny

   C. quarter quarter quarter dime penny penny penny

   D. dime penny penny

3. Rochelle bought a notebook for 62 cents. She paid with 85 cents. How much change did she get back?

   A. 13 cents
   B. 23 cents
   C. 33 cents
   D. one dollar and 47 cents

4. Joshua bought a pen for 56 cents. He paid with 75 cents. Which coins should he get back as the right change?

   A. 3 nickels, 4 pennies
   B. 1 dime, 6 pennies
   C. 2 dimes, 9 pennies
   D. 5 quarters, 1 nickel, 1 penny

5. What is the value of the coins shown?

A. 66 cents

B. 71 cents

C. 76 cents

D. 81 cents

6. Lia bought a plastic ring for 74 cents. She paid with a dollar. How much change did she get back?

A. 16 cents

B. 26 cents

C. 36 cents

D. 1 dollar and 74 cents

7. Chan bought a baseball card for 13 cents. He paid with 25 cents. Which coins should he get back as the right change?

A. 2 quarters, 3 pennies

B. 1 nickel, 7 pennies

C. 2 dimes, 7 pennies

D. 2 nickels, 2 pennies

Short-Response Questions

8. What is the value of these coins?

Answer _____

9. Jose bought a carton of chocolate milk for 69 cents. He paid with a dollar.

Part A   How much change did he get back?

Answer _____

Part B   Draw the coins he got back as change.

Lesson 35

Time

Math Words to Know

clock   a tool used to tell time

calendar   a chart used to show days of the week and months of the year

Example 1

What is the time shown on the clock?

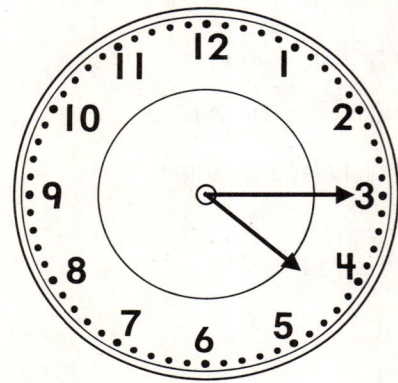

Step 1   Look at the hour hand. What is the hour?

   The hour hand is the shorter hand. It is pointing to 4.

   The hour is 4.

Step 2   Look at the minute hand. What are the minutes?

   The minute hand is the longer hand. It is pointing to 3.

   The 3 represents 15 minutes.

Hint Each number on the clock represents 5 minutes. Skip count by 5s to find the number of minutes.

Step 3   Write the hour and minutes together.

   4:15

Solution   The time shown on the clock is 4:15.

156   Measurement

Example 2

Look at the calendar. What day of the week is July 25?

JULY						
SUN	MON	TUE	WED	THU	FRI	SAT
			1	2	3	4
5	6	7	8	9	10	11
12	13	14	15	16	17	18
19	20	21	22	23	24	25
26	27	28	29	30	31	

Step 1   Find 25 on the calendar.

Step 2   Look at the top of the column where the days of the week are written.

Step 3   What is the day of the week?

Saturday

Solution   July 25 is a Saturday.

Example 3

Look at the calendar. What day is the second Monday of July?

JULY						
SUN	MON	TUE	WED	THU	FRI	SAT
			1	2	3	4
5	6	7	8	9	10	11
12	13	14	15	16	17	18
19	20	21	22	23	24	25
26	27	28	29	30	31	

Step 1   Find the Monday column.

It is the second column of the calendar.

Step 2   What is the date of the first Monday?

The first Monday is July 6.

Step 3   What is the date of the second Monday?

The second Monday is July 13.

Solution   July 13 is the second Monday of July.

Multiple-Choice Questions

1. What is the time shown on the clock?

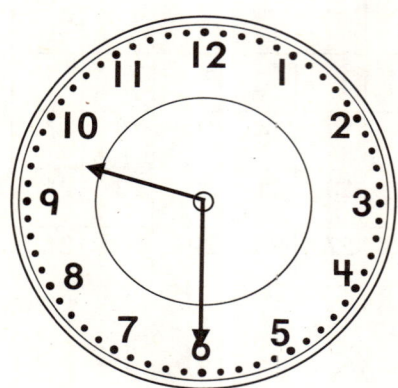

Test Taking Tip Determine the number of minutes. Cross out answers that do not have the right number of minutes.

A. 8:30   This is the right number of minutes. This could be right.

B. 9:30   This is the right number of minutes. This could be right.

C. 9:06   This is the wrong number of minutes. Cross it out.

D. 9:25   This is the wrong number of minutes. Cross it out.

2. Cathy has a meeting on September 13. What day of the week is that?

SEPTEMBER

SUN	MON	TUE	WED	THU	FRI	SAT
1	2	3	4	5	6	7
8	9	10	11	12	13	14
15	16	17	18	19	20	21
22	23	24	25	26	27	28
29	30					

A. Tuesday   C. Friday

B. Monday   D. Saturday

3. William has to visit the doctor on the third Monday of the month. What is the date of his visit?

JANUARY

SUN	MON	TUE	WED	THU	FRI	SAT
	1	2	3	4	5	6
7	8	9	10	11	12	13
14	15	16	17	18	19	20
21	22	23	24	25	26	27
28	29	30	31			

A. January 1   C. January 15

B. January 8   D. January 22

4. Mia began her homework at the time shown on the clock. What time did she begin?

A. 6:00 C. 6:25
B. 5:30 D. 6:30

5. Henry is going to a concert on June 17. What day is that?

JUNE

SUN	MON	TUE	WED	THU	FRI	SAT
					1	2
3	4	5	6	7	8	9
10	11	12	13	14	15	16
17	18	19	20	21	22	23
24	25	26	27	28	29	30

A. Sunday   C. Saturday
B. Monday   D. Tuesday

Short-Response Questions

6. Christina's birthday is on the second Tuesday of the month. What is the date of her birthday?

AUGUST

SUN	MON	TUE	WED	THU	FRI	SAT
			1	2	3	4
5	6	7	8	9	10	11
12	13	14	15	16	17	18
19	20	21	22	23	24	25
26	27	28	29	30	31	

Answer _____

7. Draw the hands on the clock to show 2:45.

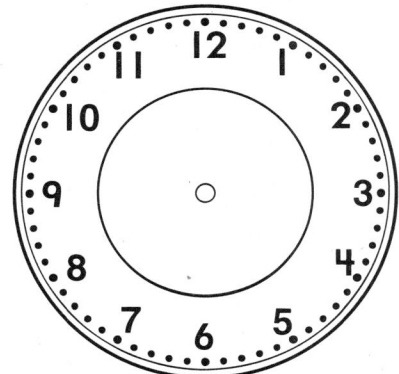

Lesson 35: Time   159

PROGRESS CHECK 1

1 Which is the **best** unit for measuring the height of a lamppost?

- A centimeter
- B liter
- C kilometer
- D yard

2 Which is the **best** unit for measuring the mass of a shoe?

- F kilogram
- G foot
- H liter
- J gram

3 Which is the **best** unit for measuring the capacity of a bathtub?

- A liter
- B kilogram
- C milliliter
- D inch

4 What is the value of the coins shown?

- F 43 cents
- G 48 cents
- H 53 cents
- J 63 cents

5 The arrow is pointing to a measurement on this inch ruler. What is the measurement?

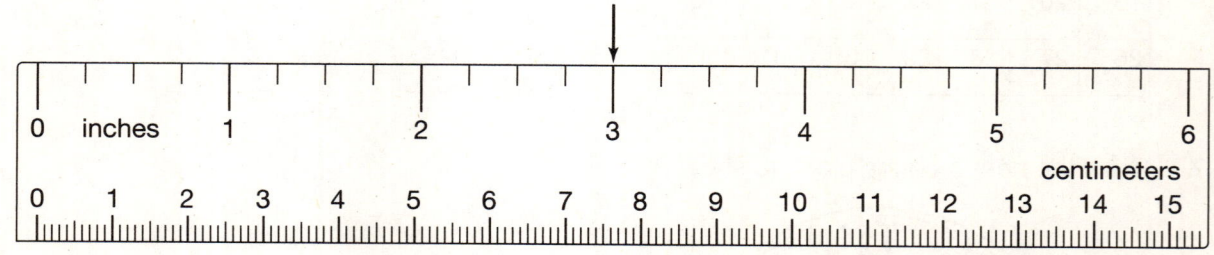

- A 2 inches
- B 3 centimeters
- C 3 inches
- D 4 centimeters

160

6 What is the time shown on the clock?

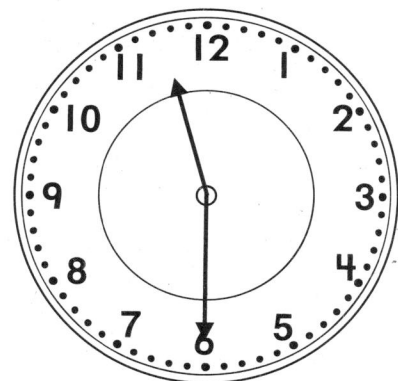

- F 11:06
- G 11:30
- H 12:30
- J 10:30

7 Which is the **best** tool for measuring the length of a grasshopper?

- A odometer
- B yard stick
- C ruler
- D meter stick

8 Peter is playing soccer on September 28. What day is that?

SEPTEMBER						
SUN	MON	TUE	WED	THU	FRI	SAT
1	2	3	4	5	6	7
8	9	10	11	12	13	14
15	16	17	18	19	20	21
22	23	24	25	26	27	28
29	30					

- F Friday
- G Sunday
- H Saturday
- J Thursday

Short-Response Question

9 Elias bought a cookie for 37 cents. He paid with a dollar.

Part A

How much change did Elias get back?

Answer _____

Part B

Write the coins Elias could have gotten back if he received correct change.

PROGRESS CHECK 2

1 What is the **best** estimate for the length of a baseball bat?

  A 3 inches
  B 3 feet
  C 3 miles
  D 3 yards

2 Which is the **best** estimate for the mass of a cow?

  F 6,000 grams
  G 6 kilograms
  H 60 kilograms
  J 600 kilograms

3 A scale shows a mass of 500 grams. Which object is most likely on the scale?

  A kitten
  B pony
  C pencil
  D car

4 What is the **best** estimate for the capacity of a juice box?

  F 25 milliliters
  G 25 liters
  H 250 milliliters
  J 2 liters

5 What is the length of this arrow to the nearest centimeter?

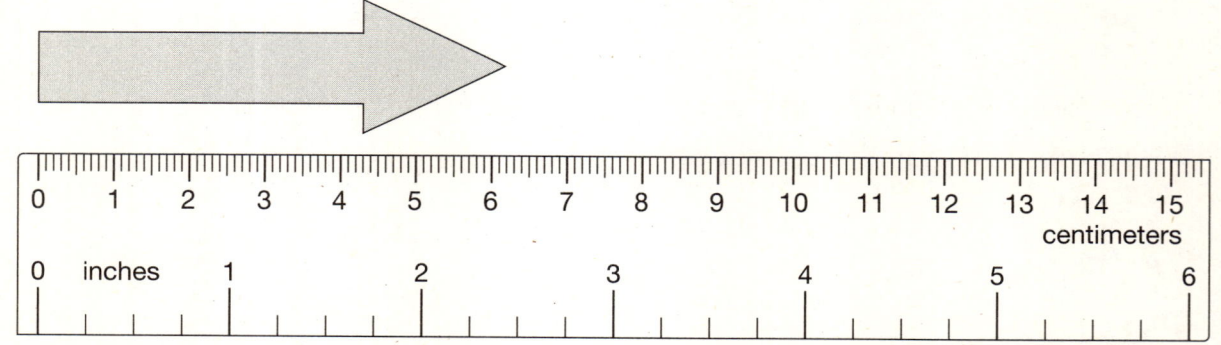

  A 5 inches
  B 5 centimeters
  C 6 inches
  D 6 centimeters

6 Marta bought a set of stickers and an album for $3.79. She paid with a $5 bill. Which choice lists her change?

    F 2 $1 bills, 1 dime, 3 nickels, 1 penny

    G 2 $1 bills, 2 dimes, 1 penny

    H 1 $1 bill, 1 dime, 2 nickels, 1 penny

    J 1 $1 bill, 1 dime, 1 penny

7 A container holds 10 milliliters. What is the container most likely to be?

    A drinking glass   **C** bucket

    B eye dropper   **D** cereal bowl

8 Johan leaves for a band trip July 17. He returns 1 week and 4 days later. What is his return date?

JULY

SUN	MON	TUE	WED	THU	FRI	SAT
			1	2	3	4
5	6	7	8	9	10	11
12	13	14	15	16	17	18
19	20	21	22	23	24	25
26	27	28	29	30	31	

    F July 27

    G July 28

    H July 29

    J July 30

Short-Response Question

9 Vanessa visited a museum from 1:30 P.M. to 4:45 P.M.

Part A

How long did she visit the museum?

Answer _____

Part B

Explain how you found your answer for Part A.

Lesson 36

Representing Data in Tables, Pictographs, and Bar Graphs

Math Words to Know

data a set of collected information

table a chart used to display data using rows and columns

graph a picture that displays data

pictograph a graph in which data is shown with symbols

bar graph a graph in which data is shown with rectangular bars

Example 1

Three students sold tickets to the school play. The table below shows the number of tickets each student sold.

Number of Tickets Sold by Students

Student	Tickets Sold
Kyle	16
Sharon	28
Vicky	19

How many more tickets did Sharon sell than Kyle?

Step 1 Look at the table. How many tickets did Sharon sell?

Sharon sold 28 tickets.

Hint The number of tickets sold is in the same row as the student's name.

Step 2 How many tickets did Kyle sell?

Kyle sold 16 tickets.

Step 3 Subtract.

28 − 16 = 12

Solution Sharon sold 12 more tickets than Kyle.

Example 2

The pictograph shows the number of CDs Salena owns.

Number of CDs Owned by Salena

Jazz	◎◎◎◎
Classical	◎◎◎
Pop	◎◎◎◎◎
Show Tunes	◎

Key
Each ◎ = 1 CD

How many jazz and classical CDs does Salena own?

Step 1 Look at the pictograph. Write the number of jazz CDs Salena owns.

She owns 4 jazz CDs.

Step 2 Write the number of classical CDs.

She owns 3 classical CDs.

Step 3 Add.

4 + 3 = 7

Solution Salena has 7 jazz and classical CDs.

Multiple-Choice Questions

Use the table to answer questions 1–3.

1. The table shows the number of points four students scored in the basketball game.

   Number of Points Scored by Students

Student	Points Scored
Kurt	13
Denise	7
Jeremy	11
Debby	14

   Who scored the most points?

   Test Taking Tip Use the table to answer the question.

A.	Kurt	Kurt scored 13 points. This could be right.
B.	Denise	Denise only scored 7 points. Cross it out.
C.	Jeremy	Jeremy only scored 11 points. Cross it out.
D.	Debby	Debby scored 14 points. This could be right.

2. How many more points did Debby score than Denise?

   A. 1
   B. 3
   C. 7
   D. 21

3. How many points did Kurt and Jeremy score in all?

   A. 18
   B. 21
   C. 24
   D. 27

166 Statistics and Probability

4. Tim collects rocks. The pictograph shows the amounts and types of rocks he has.

Number of Rocks Owned by Tim

Granite	🪨🪨
Quartz	🪨🪨🪨🪨🪨
Slate	🪨🪨🪨
Marble	🪨🪨🪨🪨🪨

Key
Each 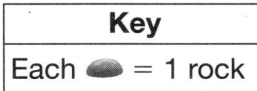 = 1 rock

How many slate rocks does Tim have?

A. 2
B. 3
C. 5
D. 6

5. The bar graph shows the number of students at school clubs one day.

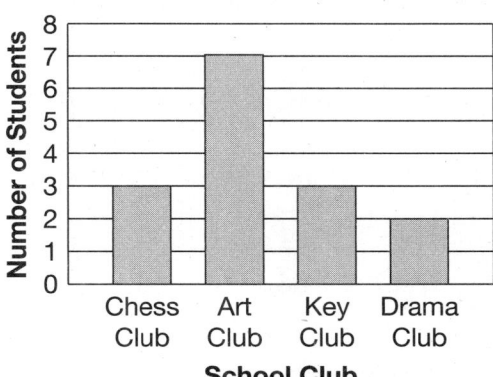

How many students were at the Art Club?

A. 2   C. 7
B. 3   D. 8

Short-Response Question

6. The school cafeteria offers four types of juices. The pictograph shows the number of cans of juice that were sold during one lunch period.

Number of Cans of Juice Sold

Key
Each 🥤 = 1 can

Part A   How many cans of orange juice were sold?

Answer _____ cans

Part B   How many more cans of cranberry juice than grape juice were sold?

Answer _____ cans

Lesson 36: Representing Data in Tables, Pictographs, and Bar Graphs   167

Lesson 37

Making Predictions from Data and Graphs

Math Words to Know

prediction a guess of what might happen based on information

Example 1

The table below shows the average price of a desktop computer at a store.

Average Price of a Desktop Computer

Year	Price
2002	$980
2004	$960
2006	$940
2008	$920

Predict the average price of a desktop computer in 2010.

Step 1 Look at the table. Are the prices increasing or decreasing?

The prices are decreasing.

Step 2 Find the pattern for the price.

Hint Use the pattern in a table or graph to make predictions.

980, 960, 940, 920

The pattern is to subtract 20 every 2 years.

Step 3 Use the pattern to predict the price of a computer in 2010.

920 − 20 = 900

Solution Prediction: the average price of a desktop computer in 2010 will be $900.

Example 2

The bar graph shows the number of miles Lakeisha walked each day.

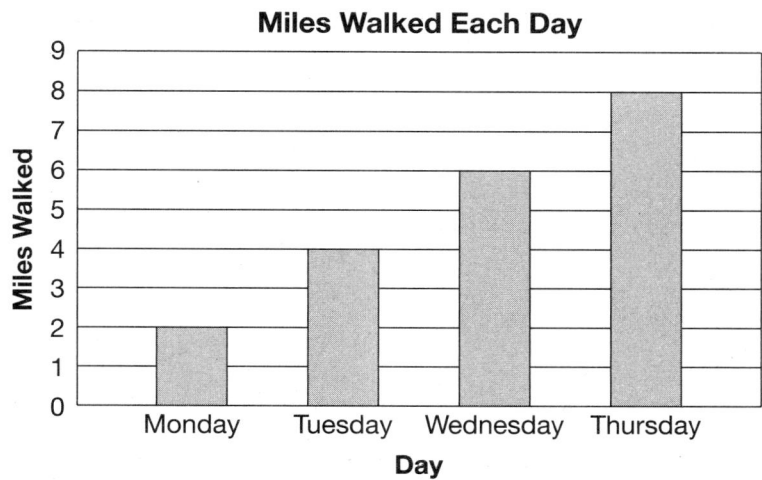

Predict the number of miles Lakeisha will walk on Friday.

Step 1 Look at the bar graph. Write the number of miles Lakeisha walked each day.

Monday: 2

Tuesday: 4

Wednesday: 6

Thursday: 8

Step 2 Write the numbers as a pattern.

2, 4, 6, 8

Step 3 Are the numbers increasing or decreasing? Write the rule.

The numbers are increasing. The rule is to add 2.

Step 4 Use the pattern to predict the number of miles Lakeisha walked on Friday.

8 + 2 = 10

Solution Prediction: Lakeisha will walk 10 miles on Friday.

Multiple-Choice Questions

1. The table shows the number of times a song was downloaded from a Web site.

   Number of Downloads of "School Days"

Week	Number of Times Downloaded
1	300
2	400
3	500
4	600

   Predict the number of times the song will be downloaded in Week 5.

   Test Taking Tip Use the table to find a pattern.

   A. 5 times — This number does not fit the pattern. Cross it out.
   B. 700 times — This number could belong in the pattern. This could be right.
   C. 750 times — This number does not fit the pattern. Cross it out.
   D. 800 times — This number could belong in the pattern. This could be right.

2. The bar graph shows the number of miles Eddie rode his bike each day.

   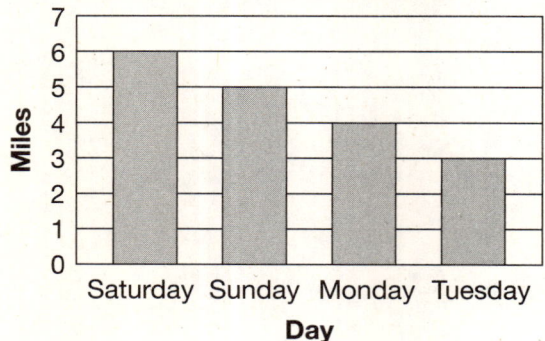

   Predict the number of miles Eddie will ride his bike on Wednesday.

   A. 0 miles
   B. 1 mile
   C. 2 miles
   D. 3 miles

170 Statistics and Probability

3. The table shows the number of state capitals Alex has memorized.

Number of State Capitals Memorized by Alex

Day	Number of State Capitals
Tuesday	8
Wednesday	14
Thursday	20
Friday	26

Predict the number of state capitals Alex will have memorized by Saturday.

A. 28 state capitals

B. 30 state capitals

C. 32 state capitals

D. 34 state capitals

4. The bar graph shows the number of laps Nicki swam each day.

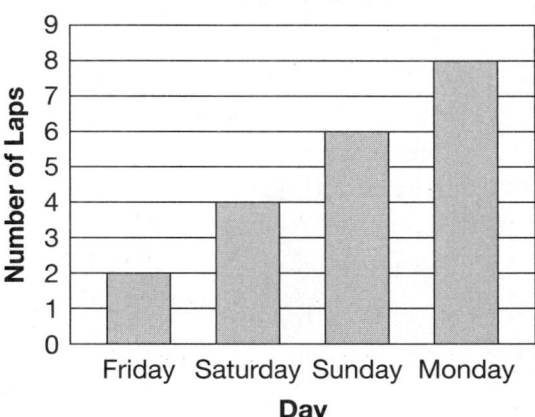

Predict the number of laps Nicki will swim on Tuesday.

A. 6 laps C. 9 laps

B. 8 laps D. 10 laps

Short-Response Question

5. Xavier goes fishing every week. The pictograph shows the number of fish he caught each week.

Number of Fish Caught by Xavier

Week 5	🐟 🐟 🐟 🐟 🐟 🐟
Week 6	🐟 🐟 🐟 🐟 🐟
Week 7	🐟 🐟 🐟 🐟
Week 8	🐟 🐟 🐟

Key
Each 🐟 = 1 fish

Predict the number of fish Xavier will catch in Week 9.

Answer _____ fish

Lesson 37: Making Predictions from Data and Graphs

PROGRESS CHECK 1

Use the table to answer questions 1–3.

The table shows the number of questions students got right on a test.

Number of Questions Answered Right

Student	Number of Right Answers
Shawn	88
Nicole	95
Felix	79
Randi	93

1 Who got the most questions right?

  A Shawn
  B Nicole
  C Felix
  D Randi

2 How many more questions did Randi get right than Felix?

  F 5
  G 7
  H 14
  J 16

3 How many questions did Shawn and Nicole get right in all?

  A 7
  B 173
  C 181
  D 183

Use the bar graph to answer questions 4–6.

The bar graph shows the number of pets owned by students.

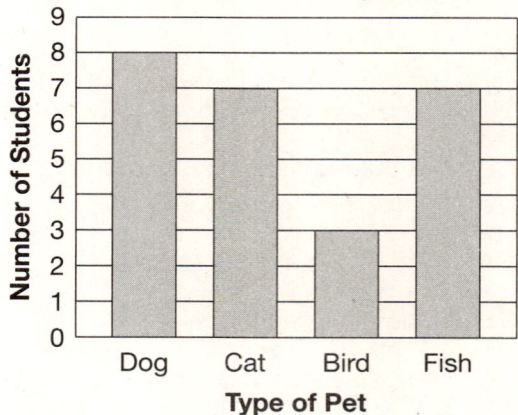

4 How many students own a dog?

  F 3
  G 7
  H 8
  J 9

5 Which two kinds of pets are owned by the same number of students?

  A cats and fish
  B dogs and cats
  C dogs and fish
  D cats and birds

6 How many more students own a cat than a bird?

  F 1
  G 2
  H 3
  J 4

7 The table shows the number of people that attended the school play.

School Play Attendance

Performance	Number of People
1	230
2	210
3	190
4	170

Predict the number of people that will attend the play at the next performance.

A 150   **C** 170

B 160   **D** 180

8 The pictograph shows the number of comic books Luis read each week.

Number of Comic Books Read by Luis

Week 2	📖
Week 3	📖📖📖
Week 4	📖📖📖📖📖
Week 5	📖📖📖📖📖📖📖

Key
Each 📖 = 1 comic book

Predict the number of comic books Luis will read in Week 6.

F 6   **H** 8

G 7   **J** 9

Short-Response Question

9 The table shows the high temperatures in Rockland County for four days.

Rockland County High Temperature

Day	Temperature (°F)
Wednesday	77
Thursday	74
Friday	71
Saturday	68

Part A

Which day was the warmest?

Answer _____

Part B

How many more degrees warmer was it on Thursday than Saturday?

Answer _____ °F

Part C

Predict the temperature for Sunday.

Answer _____ °F

PROGRESS CHECK 2

Use the pictograph to answer questions 1–3.

The pictograph shows the number of model cars owned by four students.

Number of Model Cars Owned

Ron	🚗 🚗 🚗
Marisa	🚗
Kenny	🚗 🚗 🚗 🚗 🚗 🚗
Tonya	🚗 🚗 🚗 🚗

Key
Each 🚗 = 1 car

1 How many model cars does Ron own?

A 1   C 4
B 3   D 6

2 How many model cars do Marisa, Ron, and Tonya own in all?

F 4   H 8
G 7   J 9

3 How many more model cars does Kenny own than Tonya?

A 2   C 4
B 3   D 10

4 The bar graph shows the number of different cities Carlos visited.

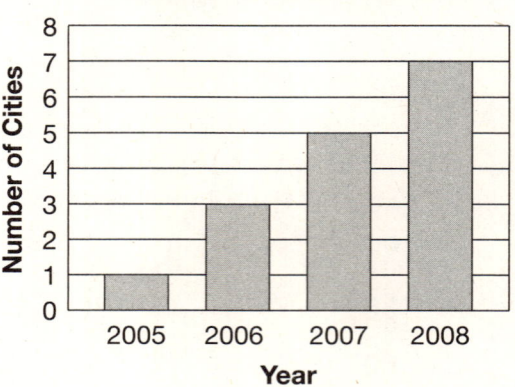

Predict the number of cities Carlos will visit in 2009.

F 7   H 9
G 8   J 10

5 The bar graph shows the number of hours Michelle went snowboarding.

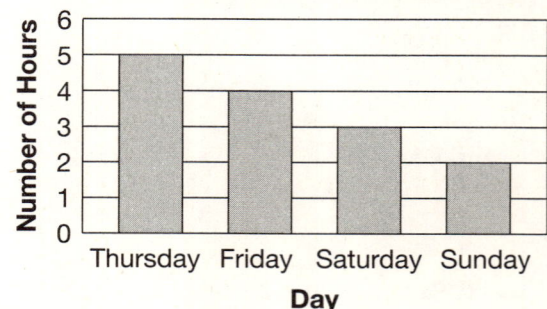

Predict the number of hours Michelle will snowboard on Monday.

A 0   C 2
B 1   D 6

6 The table shows the number of phone minutes used by the Suzuki family each month.

Phone Minutes Used

Month	Number of Minutes
March	870
April	840
May	810
June	780

Predict the number of minutes the Suzuki family will use the phone in July.

F 750   H 770

G 760   J 790

7 The pictograph shows the number of bracelets Janine made.

Number of Bracelets Made

Monday	◯◯◯
Tuesday	◯◯◯◯
Wednesday	◯◯◯◯◯
Thursday	◯◯◯◯◯◯

Key
Each ◯ = 1 bracelet

Predict the number of bracelets Janine will make on Friday.

A 6   C 8

B 7   D 9

Short-Response Question

8 This graph shows the height of a tree every 10 years.

Part A

Predict the height of the tree after 90 and 100 years.

Answer _____

Part B

Explain how you found your answer for Part A.

Tree Height in Inches

Progress Check 2   175

New York State Progress Coach,
Mathematics, Grade 4

PROGRESS TEST

Name: _____

TIPS FOR TAKING THE TEST

Here are some suggestions to help you do your best:

- Be sure to read carefully all the directions in the test book.
- You may use the Punch-Out Tools on p. 201 to help you answer questions on the Progress Test.
- Read each question carefully and think about the answer before choosing your response.

 This picture means that you will use your ruler.

Session 1

1 Which list is in order from greatest to least?

　A　84, 824, 834, 843

　B　843, 824, 834, 84

　C　84, 843, 834, 824

　D　843, 834, 824, 84

Hint Compare the place value of the digits.

2 Which is 358 rounded to the nearest hundred?

　F　300

　G　350

　H　360

　J　400

3 What is the **best** estimate for the capacity of a sports bottle?

　A　100 milliliters

　B　1 liter

　C　10 liters

　D　100 liters

Hint Picture how much water is in a sports bottle.

4 Which expression has the same value as (8 × 2) × 5?

　F　8 × (2 × 5)

　G　(8 + 2) + 5

　H　(8 × 2) + 5

　J　8 + (2 × 5)

Hint Remember the Associative Property of Multiplication.

5 Which number can you multiply by 87 so the product is an odd number?

　A　16

　B　30

　C　87

　D　94

6 Which pattern shows *add 3*?

F
Input	7	8	9	10	11
Output	10	11	12	13	14

G
Input	21	18	15	12	9
Output	7	6	5	4	3

H
Input	5	6	7	8	9
Output	15	18	21	24	27

I
Input	1	2	3	4	5
Output	3	9	27	81	243

Hint *Add 3* to each input number.

7 48 × 7 = _____

A 326

B 336

C 346

D 356

Hint Write the problem vertically.

8 Which number makes this open sentence true?

38 > _____

F 37

G 38

H 39

J 40

9 Eduardo ate $\frac{1}{4}$ of a pie. What is another name for this fraction?

A $\frac{3}{4}$

B $\frac{1}{2}$

C $\frac{4}{10}$

D $\frac{5}{20}$

Hint Multiply the numerator and denominator by the same factor.

Go On

10 Naomi skip counted by 1,000. She started at 1,000. Which number did Naomi say next?

F 2,001

G 2,100

H 2,000

J 3,000

11 What is the area of this rectangle?

A 24 square units

B 30 square units

C 32 square units

D 36 square units

Hint Count the number of square units in the rectangle.

12 A number has 8 ones, 3 tens, 4 thousands. What is the number?

F 438

G 834

H 4,038

J 4,308

13 Which number sentence is represented by this model?

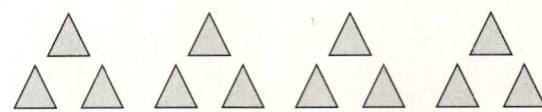

A 4 × 3 = _____

B 3 + 4 = _____

C 4 ÷ 3 = _____

D 4 − 3 = _____

Hint Count the number of triangles in each group. Count the number of groups. What operation is shown?

14 A class contest begins on September 6 and ends on September 30. How many weeks and days does the contest last?

SEPTEMBER

SUN	MON	TUE	WED	THU	FRI	SAT
1	2	3	4	5	6	7
8	9	10	11	12	13	14
15	16	17	18	19	20	21
22	23	24	25	26	27	28
29	30					

F 3 weeks 1 day

G 3 weeks 4 days

H 3 weeks 5 days

J 4 weeks 2 days

Hint Be sure to count the first and last days.

15 What is the value of the 6 in 8,630?

A 6

B 60

C 600

D 6,000

16 The table shows the number of Spanish words Tony memorized. What is the best prediction for the number of words he will memorize on Sunday?

Number of Spanish Words Tony Memorized

Day	Words Memorized
Wednesday	35
Thursday	42
Friday	49
Saturday	56

F 59   H 62

G 61   J 63

Hint Find the pattern for the number of words memorized each day.

17 Which tool would be best to measure the driving distance from Binghamton to New York City?

A odometer   C ruler

B scale   D yard stick

Hint Cross out the tool that is not used for measuring length.

18 If these numbers were ordered from least to greatest, which number would be third on the list?

8,496   8,501   8,494   8,512

F 8,494   H 8,501

G 8,496   J 8,512

Go On

19 Which symbol shown below makes this sentence true?

54 _____ 45

A <
B =
C –
D ≠

20 Tasha wants to find out the favorite TV show of students in her class. What question should she ask her classmates?

F Do you watch TV?
G What is your favorite TV show?
H Is *Friends for Life* your favorite TV show?
J What TV shows do you watch?

21 Which is a true statement?

A $\frac{1}{3} > \frac{1}{2}$
B $\frac{1}{4} < \frac{1}{3}$
C $\frac{1}{2} < \frac{1}{6}$
D $\frac{1}{3} = \frac{1}{6}$

22 Which figure is congruent to the one below?

F
G
H
J

Hint Which rectangles have the same size **and** shape?

23 Mr. Chi is writing an order for 1,806 bricks. How do you write the word name for this number?

A one thousand, eight six
B one thousand, eighty-six
C eighteen thousand, six
D one thousand, eight hundred six

24 Which number sentence belongs to the same fact family as 48 ÷ 8 = 6?

F 6 × 8 = 48

G 48 ÷ 12 = 4

H 48 − 6 = 42

J 8 ÷ 4 = 2

Hint Fact families have the same numbers for multiplication and division facts.

25 What fraction does the letter *X* represent on this number line?

A $\frac{1}{6}$ C $\frac{1}{5}$

B $\frac{1}{4}$ D $\frac{1}{7}$

26 Which of these has the least product?

F 20 × 1

G 6 × 3

H 4 × 7

J 5 × 5

Hint Multiply. Then compare the products.

27 A doorway measures 3 yards high. Jasmine is making a door that is 2 inches shorter than the height of the doorway. What is the height of the door that Jasmine is making?

A 2 feet, 10 inches

B 8 feet

C 8 feet, 8 inches

D 8 feet, 10 inches

Hint Convert the height of the doorway from yards to feet.

28 Which open sentence can be used to solve this problem?

Brent has a pack of gum with 20 pieces. He gives five pieces to his friends. How many pieces of gum does he have left?

F 20 ÷ 5 = _____

G 20 ÷ 4 = _____

H 20 − 5 = _____

J 20 + 5 = _____

Go On

29 For which situation can you use an estimate?

- A finding the amount of change you get back after paying a bill with $50
- B finding if you have enough money to pay for items in a shopping cart
- C finding how much it costs for two people to ride the bus
- D finding how much you earn each hour at a job

Hint Which situation does not require an exact amount?

30 A scale reads 100 grams. Which object is most likely on the scale?

- F pencil
- G car
- H bicycle
- J calculator

STOP

Session 2

31 Kelly drew a hexagon.

Part A

How many angles make up the figure?

Answer _____ angles

Part B

Explain how you found your answer for Part A.

_____

_____

Hint A hexagon has the same number of angles as the number of sides.

32 Hailey bought 56 hamburger buns for a picnic. The buns come in packages of 8. How many packages of hamburger buns did Hailey buy?

Show your work.

Answer _____ packages

Hint Use division.

Go On

33 Scott bought an eraser for $0.37. He paid with a $1 bill.

Part A

List the coins Scott might get back as change.

Answer _____

Part B

On the lines below, explain how you got your answer.

_____
_____
_____
_____

Hint Subtract $1.00 − $0.37. Be sure to line up the decimal points.

34 Jackie read a book that was 346 pages.

Part A

How many pages did she read, rounded to the nearest hundred?

Answer _____ pages

Part B

Explain how you got your answer in Part A.

_____
_____
_____
_____

Hint Look at the tens digit. Is it less than, equal to, or greater than 5?

35 Craig is making a window decoration in the shape of a regular pentagon. What is the perimeter of the decoration?

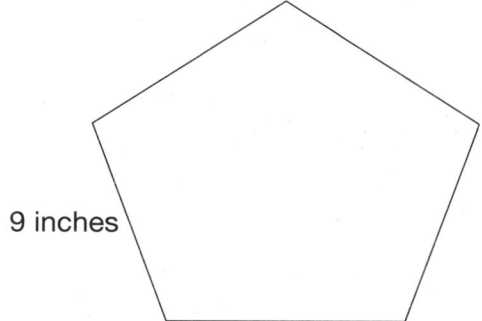

9 inches

Show your work.

Answer _____ inches

Hint A regular pentagon has sides that all measure the same length.

36 This list shows the number of centimeters of snow on a snow bank over 4 days.

2, 6, 18, 54,…

Part A

If the pattern continues, how much snow will be on the snow bank after the fifth day?

Answer _____ centimeters

Part B

If the pattern continues, how much snow will be on the snow bank after the sixth day?

Answer _____ centimeters

Hint Find the rule to the pattern. Then use the rule to solve the problem.

Go On

37 The school cafeteria sold 4,293 cartons of chocolate milk in March. They also sold 1,967 cartons of plain milk. The cafeteria manager said they sold a total of 7,260 cartons of plain milk and chocolate milk for the month.

Part A

Is the manager's answer reasonable? Use estimation to determine your answer.

Answer _____

Part B

Explain how you used estimation to get your answer in Part A.

_____
_____
_____

38 Candace is packing 74 light bulbs into boxes. Six light bulbs fit in one box.

Part A

How many boxes will she need?

Answer _____ boxes

Part B

Explain how you got your answer in Part A.

_____
_____
_____

39 John bought a poster. The poster was shipped in a box in this shape.

Part A

What three-dimensional figure is the box?

Answer _____

Hint What shapes are the faces of the figure? Use the shapes of the faces to help determine the figure.

Part B

Explain how you found your answer. Use the words *edge, face,* and *vertices* in your answer.

_____

_____

_____

_____

Part C

Name another three-dimensional figure that has triangular faces.

Answer _____

Session 3

40 This bar graph shows the results of a survey asking 4th graders about their favorite New York City attraction.

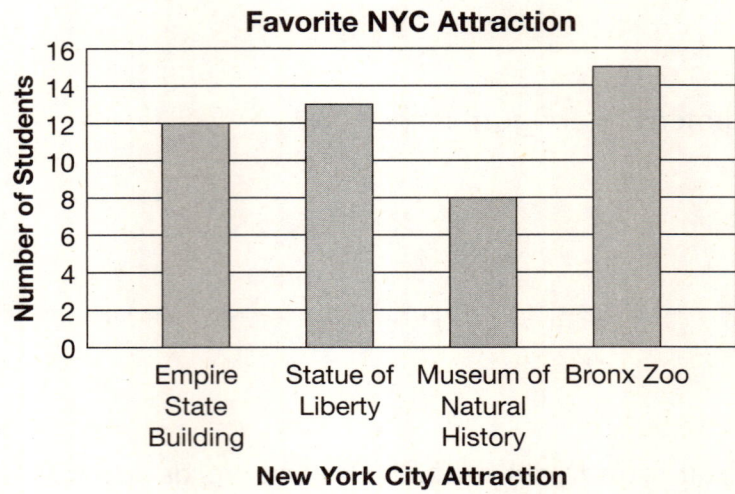

Part A

How many more students voted for the Bronx Zoo than the Museum of Natural History?

Answer _____ students

Part B

Explain how you found your answer for Part A.

_____

_____

_____

_____

Hint Use the bar graph to find the number of votes for the Bronx Zoo compared to votes for the Museum of Natural History. Which operation will you use?

41 Trina goes to school from 7:30 A.M. to 3 P.M. For how long does she go to school?

Show your work.

Answer _____

42 A jar holds 800 jellybeans. A store sold 300 jars.

Part A

Find the total number of jellybeans the store sold.

Show your work.

Answer _____ jellybeans

Hint Which operation will you use to find the total number in 300 groups?

Part B

Explain how you found your answer in Part A.

_____

_____

Go On

43 Michael is keeping a worm in a box with dirt.

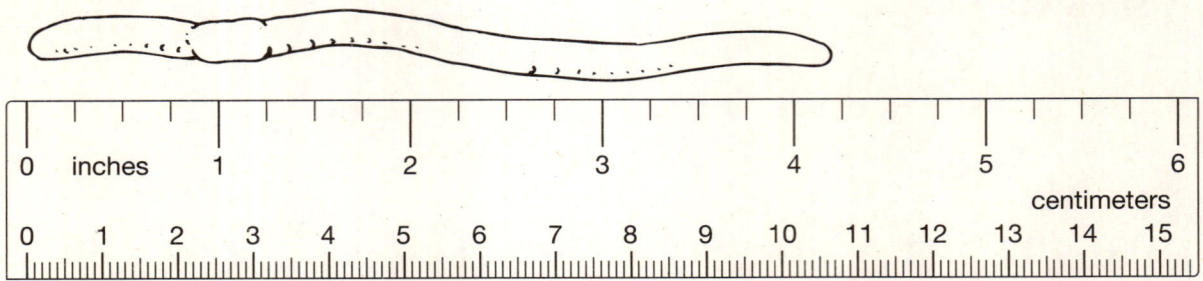

Part A

What is the length of the worm to the nearest quarter inch?

Answer _____ inches

Hint Write fractions below each quarter-inch mark on the ruler.

Part B

What is the length of the worm to the nearest inch?

Answer _____ inches

44 The 4th grade class raised $3,226 for charity. The 5th grade class raised $2,759 for charity. How much more money did the 4th grade class raise than the 5th grade class?

Show your work.

Answer $ _____

Hint Which operation will you use to find how much more? Write the problem vertically.

45 Theresa saw that a picture was in this shape.

Part A

What is the name for this figure?

Answer _____

Hint Count the number of sides on the figure.

Part B

Explain how you got your answer in Part A.

_____

_____

Go On

Session 3

Progress Test 193

46 Trey plays chess for 4 hours each day. He played chess for 9 days. How many hours did Trey play chess in all?

Show your work.

Answer _____ hours

Hint Which operation will you use?

47 Latisha made this input-output table.

Input	Output
20	5
28	7
44	
64	16

Part A

What is the missing output number?

Answer _____

Hint First find the rule for the input-output table. Use the rule to find the missing output number.

Part B

If the next output number is 24, what number belongs in the input column?

Answer _____

48 Dwayne asked the students in his class a survey question. Then he made this list.

volleyball	soccer	football
football	soccer	baseball
soccer	volleyball	football
baseball	baseball	soccer
football	soccer	soccer
volleyball	volleyball	football

Part A

What survey question do you think Dwayne asked?

Answer _____

Part B

Make a tally chart of Dwayne's data.

Part C

How many more students voted for soccer than baseball?

Answer _____ students

Hint In your tally chart, have one tally mark equal one vote.

STOP

Glossary

A

addition combining groups to find the total amount (page 60)

angle a figure that is made up of two rays or line segments with the same endpoint (page 120)

area the number of square units needed to cover a figure (page 124)

array pictures, numbers, or objects arranged in rows and columns (page 12)

Associative Property of Multiplication a rule that states numbers grouped in any way will have the same product: $(7 \times 5) \times 2 = 7 \times (5 \times 2)$ (page 52)

B

bar graph a graph in which data is shown by rectangular bars (page 164)

C

calendar a chart used to show days of the week and months of the year (page 156)

capacity the measure of how much a container can hold (page 148)

clock a tool used to tell time (page 156)

closed figure a shape with no openings or gaps (page 120)

column a line in a table or an array that goes up and down (pages 16 and 64)

cone a pointed, three-dimensional figure with a circular base (page 128)

congruent figures plane figures that have the same size and the same shape (page 32)

counting number 1, 2, 3, etc. (page 60)

cube a three-dimensional figure with faces that are squares (page 128)

customary system a measurement system using inches, feet, and yards to measure length (page 140)

cylinder a three-dimensional figure with matching faces that are circles; the figure has no edges or vertices (page 128)

D

data a set of collected information (pages 36 and 164)

denominator the bottom number in a fraction, such as the 2 in $\frac{1}{2}$ (page 4)

dividend in division, the number that is divided (pages 12 and 80)

division separation of amounts into smaller, equal groups (page 12)

division fact a division sentence, such as $8 \div 2 = 4$ (page 68)

divisor in division, the number that divides the dividend (pages 12 and 80)

dollar a unit of money equal to 100 cents (page 152)

E

edge a line segment where two faces meet on a three-dimensional figure (page 128)

equivalent fractions different fractions that have equal value, such as $\frac{1}{2}$ and $\frac{2}{4}$ (page 4)

estimate a number that is close to the exact amount (pages 24 and 92)

even number a number you can divide evenly by 2; the ones digit of an even number is always 0, 2, 4, 6, or 8 (page 56)

expanded form a way of writing a number as the sum of the values of its digits; 3,762 in expanded form is $3,000 + 700 + 60 + 2$ (page 44)

F

face the flat side of a three-dimensional figure (page 128)

fact family facts that are related, using the same numbers (page 68)

G

graph   a picture that displays data   (page 164)

H

hexagon   a polygon with 6 sides and 6 angles   (page 120)

I

input-output table   a table that uses a rule to change each input number to an output number   (page 112)

is greater than (>)   an amount that is more than a second amount   (pages 8, 28, and 48)

is less than (<)   an amount that is less than a second amount   (pages 8, 28, and 48)

L

length   the distance from one point to another   (page 136)

like denominators   fractions with the same denominator, such as $\frac{2}{4}$ and $\frac{3}{4}$ (page 8)

line segment   a part of a line that has two endpoints   (page 120)

M

mass   the amount of matter in an object   (page 144)

measure   to find the size, or capacity of an object   (page 136)

metric system   a measurement system using millimeters, centimeters, meters, and kilometers to measure length   (page 140)

multiple   the product of a number and any whole number   (page 72)

multiplication fact   a multiplication sentence, such as $4 \times 2 = 8$ (page 68)

multiplication table   a table of multiplication facts   (page 16)

N

not equal to   shown with the symbol $\neq$ (page 104)

number sentence   a math sentence written with numbers, operation signs ($+$, $-$, $\div$, $\times$), and comparing symbols ($<$, $>$, $=$)   (page 100)

numerator   the top number in a fraction, such as the 1 in $\frac{1}{2}$ (page 4)

O

octagon   a polygon with 8 sides and 8 angles   (page 120)

odd number   a number you cannot divide evenly by 2; the ones digit of an odd number is always 1, 3, 5, 7, or 9   (page 56)

open sentence   a number sentence that is missing one number   (page 100)

operation   addition, subtraction, multiplication, or division   (page 84)

P

pattern   a series of numbers or shapes that follow a rule   (page 108)

pattern rule   an operation performed on numbers, or an action performed on shapes, in a pattern   (page 108)

pentagon   a polygon with 5 sides and 5 angles   (page 120)

perimeter   the measure of the distance around a figure   (page 124)

pictograph   a graph in which data is shown with symbols   (page 164)

place value   the value of a digit in a number   (page 44)

plane figure   a flat, two-dimensional figure like a square, circle, or rectangle   (page 32)

polygon   a closed figure made up of three or more line segments   (page 120)

prediction   a guess of what might happen based on information   (page 168)

prism   a three-dimensional figure with matching faces that are polygons   (page 128)

product   the answer in a multiplication problem   (pages 16 and 76)

pyramid   a three-dimensional figure with a base that is a polygon; the other faces are triangles   (page 128)

Q

quadrilateral   a polygon with 4 sides and 4 angles; examples include a square, rectangle, parallelogram, and rhombus   (page 120)

quotient   the answer to a division problem   (pages 12 and 80)

R

ray   part of a line that has one endpoint and goes on forever in one direction   (page 120)

rounding   to estimate the value of a number based on a given place value   (pages 20 and 88)

row   a line in a table or an array that goes across   (pages 16 and 64)

S

similar figures   plane figures that have the same shape but, not necessarily the same size   (page 32)

sphere   a three-dimensional figure with no faces, edges, or vertices   (page 128)

standard form   a number written with only numerals, such as 3,762   (page 44)

subtraction   taking away to find a difference   (page 60)

survey   information about a group that is found by questioning or observing   (page 36)

T

table   a chart used to display data using rows and columns   (page 164)

tally table   a table that shows data by using a tally mark for each item counted   (page 36)

triangle   a polygon with 3 sides and 3 angles   (page 120)

U

unit fraction   a fraction with a numerator of 1, such as $\frac{1}{2}$ (pages 4 and 28)

V

vertex   a point where three or more edges meet on a three-dimensional figure   (page 128)

W

whole number   a number that is not a fraction or decimal (0, 1, 2, 3, etc.)   (page 60)

Punch-Out Tools

Photocopy this page and cut out the tools.

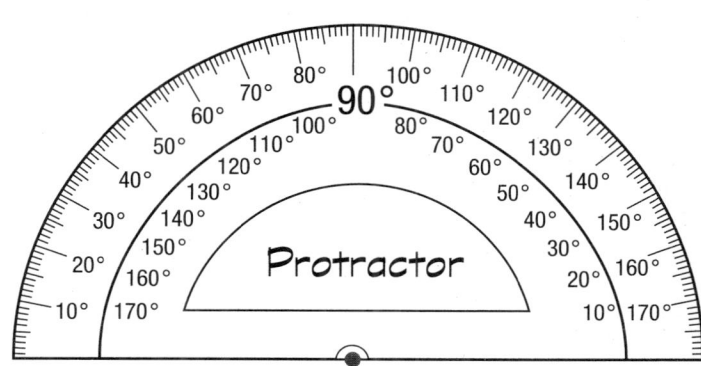

Ruler

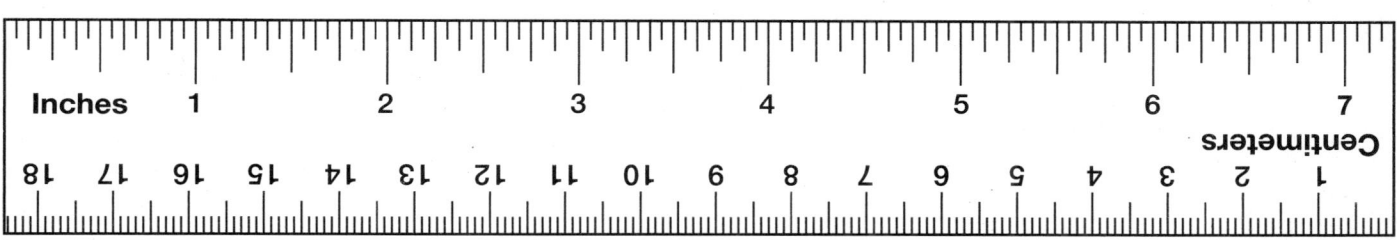

Notes

Notes

Notes

Notes

Notes

Notes